HERBIVORY

Edited by **Breno Barros** and **Marcus E. B. Fernandes**

Herbivory

http://dx.doi.org/10.5772/2718
Edited by Breno Barros and Marcus E. B. Fernandes

Contributors

Allan Sebata, Rita de Cassia Oliveira dos Santos, Marcus Emanuel Barroncas Fernandes, Marlucia Bonifácio Martins, Breno Barros, Yoichi Sakai, Hiroaki Hashimoto, Kenji Gushima, Yrlan Oliveira, Fernando Araújo Abrunhosa, Marcelo Vallinoto, Michael J. Stout, Roger S. Ingram, Morgan P. Doran, Glenn Nader

Published by InTech

Janeza Trdine 9, 51000 Rijeka, Croatia

Publishing Process Manager Sandra Bakic
Typesetting InTech DTP Team
Cover InTech Design Team

Additional hard copies can be obtained from orders@intechopen.com

Herbivory, Edited by Breno Barros and Marcus E. B. Fernandes
p. cm.
ISBN 978-953-51-1052-1

For EU product safety concerns:
IN TECH d.o.o., Prolaz Marije Krucifikse Kozulić 3, 51000 Rijeka, Croatia,
info@intechopen.com or visit our website at intechopen.com.

Meet the editors

Dr. Breno Barros graduated as B.Sc. in Biological Sciences (Universidade Federal do Pará, Brazil) in 2002. He attended the Hiroshima University (Japan) in 2005, where he has achieved his M.Sc. degree in Marine Ecology at the Laboratory of Aquatic Resources (2007). In 2012, he has concluded his Ph.D. in Environmental Biology at the Institute of Coastal Studies (IECOS), Federal University of Pará, Brazil, with a brief sandwich period in the Laboratory of Aquatic Resources, Hiroshima University, Japan. Dr. Barros has been studying teleost fish feeding behavior, focusing his attentions to the Family Ephippidae, establishing comparative studies on behavioral ecology of Ephippid fish from both Pacific and Atlantic oceans, having published a number of papers dealing with feeding behavior and mimicry by representatives of that group. Currently, he is a post-doctorate candidate at the Laboratory of Biological Systematics and Co-evolution at the Institute of Coastal Studies (IECOS), Federal University of Pará, Brazil.

Dr. Marcus E. B. Fernandes graduated from the Federal University of Amazonas, Brazil, in 1985, before going to the University of São Paulo, to take his Master degrees in 1989. Between 1989 and 1993, he was at the Department of Zoology, Goeldi Museum in Belém, before going on to receive his DPhil from the University of York, England, in 1998. Marcus then lectured ecology at the Federal University of Maranhão, for four years as a visiting professor. In 2002, he returned to the Amazon region to teach ecology at the Federal University of Pará, where he is currently a tenured professor. He is also a consultant for a number of Brazilian and international research funding agencies, and is a reviewer for a range of journals. He has written or co-authored more than 40 research papers and has authored / edited five books, and more than 10 book chapters and numerous magazine and newspaper articles on scientific subjects. He is currently involved in research on the invertebrate fauna associated with mangrove tree species and the effects of herbivory on mangrove restoration.

Contents

Preface

Plant and animal species have been co-evolving synergically ever since they began to occupy the different environments found around the world. These two types of organisms have developed relationships of mutual dependence at a number of different levels, in particular those related to feeding behavior. The study of herbivore interactions has been especially relevant to the scientific community. This interspecific relationship is present in most known biological systems, and is especially relevant from both evolutionary and applied viewpoints, given the potential for the understanding of ecological phenomena ranging from the simplest to the most complex processes. Relationships of this kind are based on a cost-benefit dichotomy, which may be expressed in many different forms, ranging from mutual benefits for both parties, such as the control of plant growth by wild ruminants, to the parasitic relationships observed in many insect-plant interactions. It is important to remember, however, that any event that may cause disequilibrium in any kind of herbivorous activity, such as changes in the environment, anthropogenic impacts, species extinctions or bio-invasions, may provoke significant changes not only to the species involved in the interaction, but eventually to the system as a whole.

Studies of herbivory provide important insights into fundamental questions in many areas of the biological sciences. The focus on natural systems is more effective for the prediction of potential changes in ecosystems, given that agricultural systems are designed to create an equilibrium that optimizes the productive process. Given the ramifications of the processes related to herbivory, studies based on complementary approaches are necessary for a better understanding of the different aspects of the ecological process. This book attempts to expand on these different aspects of herbivory by presenting a multidisciplinary approach to a number of different themes, focusing on topics that range from basic research in natural habitats to the intrinsic relationships between animals and plants in agricultural systems.

The relationship between herbivores and woody plants in semi-arid savanna ecosystems will be examined, together with an analysis of the key adaptations which allow plants to persist in these environments under intense pressure from herbivores. In the humid zone, herbivory is represented by a review of the association between the black mangrove (genus *Avicennia*) and gall-inducing insects, testing the hypothesis that the species of this genus may be "superhosts" in mangrove forests throughout the world. Herbivory was also analyzed in the aquatic environment, through a comparative study of the role of ephippid fishes in the coastal waters of Brazil and Japan, where these species are considered to have an important functional role in the ecosystem. In addition to these natural environments, herbivory is also discussed in relation to agricultural and livestock rearing activities. An initial approach reiterates the importance of the herbivore process, in terms of the resistance of plants in both wild and agricultural environments. The second and final contribution evaluates the role of planned herbivory, that is, the use of grazing animals as a strategy for the prevention of wildfires in natural environments.

Dr. Marcus E. B. Fernandes
Universidade Federal do Pará,
Brazil

Dr. Breno Barros
Universidade Federal do Pará, Instituto de Estudos Costeiros,
Brazil

Chapter 1

Woody Plant-Herbivore Interactions in Semi-Arid Savanna Ecosystems

Allan Sebata

Additional information is available at the end of the chapter

http://dx.doi.org/10.5772/48400

1. Introduction

Savannas cover more than ten percent of the world's land surface and more than fifty percent of Africa, providing browse to millions of mammalian herbivores (Scogings & Mopipi, 2008). Although herbivory is a major driver of ecosystem functioning in semi-arid African savannas plant-herbivore interactions are poorly understood (Skarpe, 1992; Scholes, 1997; Scogings, 2003). African savannas and large herbivores coevolved, with woody plants developing defences against herbivory (Du Toit, 2003). The herbivores have in turn evolved counter measures against the plant defences. Large herbivores counteract the effects of plant defence by selective foraging, fragmentation of intact plant tissues, microbial fermentation and expanded guts for microbial breakdown, whereas plants protect themselves through morphological, structural and chemical adaptations (Borchard et al., 2011). African savanna ecosystems under heavy browsing have few hardy woody species that are resistant to or are defended against defoliation. Cornell and Hawkins (2003) suggested that plants acquire better defences with time which herbivores in turn learn to partly or fully overcome. Hartley & Jones (1997) found woody plants to be able to live in environments where herbivores were common because of their ability to resist or recover from intense herbivore pressure. The varying defences that plants exhibit is a reflection of the diversity of herbivores and abiotic conditions. Plant defences exert selective pressure on mammalian herbivores, with the result that many have developed mouthparts and digestive systems that facilitate the use of particular plant types. The chemical defences of terrestrial plants reflects in part the biochemical evolution of early land plants and the problems those plants encountered.

A number of plant defence theories have been advanced to explain why some plants are better defended than others. For example, the optimal defence hypothesis focuses on how defensive needs of plants leads to the evolution of chemical defences, with the cost of that defence maximizing fitness. This chapter will discuss the effects of herbivory on woody plants, show how the plants respond to herbivory and explore herbivore adaptations to plant defences. I will also discuss the woody plant-herbivore interactions in terms of browse instantaneous intake rates and explain how shoot morphology influences herbivory.

2. Effects of herbivory on woody plants

Herbivory can negatively through instantaneous death (Belsky, 1986) or positively through increased growth and competitive ability (McNaughton, 1979) influence plant fitness. The effect of herbivory on woody plants depend on the intensity and frequency of damage, plant phenological stage and resource relationships at the time of herbivory, plant tissues removed, competition with non-browsed species and the characteristics of the plant species (Maschinski & Whitham, 1989). Damage to individual woody plant branches negatively affects growth and reproduction of those branches but leads to compensatory growth in non damaged branches (Du Toit et al., 1990). Many woody species in the semi-arid savanna are able to resprout following herbivory. For example, *Acacia karroo* has the ability to coppice strongly following defoliation (Teague & Walker, 1988). Herbivory stimulates shoot production in mature *Acacia* trees (Dangerfield & Modukanele, 1996) and root growth in *Faidherbia albida* (Dube et al., 2009), while negatively affecting *Acacia* seedling growth (Walker, 1985). The resprouting of woody plants after damage by herbivores depends on their ability to use stored nutrient reserves and on the buds that escaped herbivory and can be activated for new growth. Woody plants differ in their ability to recover after herbivory with resprouting being influenced by the rate of regrowth of leaves and shoots. In a study to compare the compensatory abilities of three *Acacia* species, *Acacia karroo* fully compensated while *Acacia nilotica* overcompensated with *Acacia rehmanniana* under-compensating lost biomass (See Figure 1).

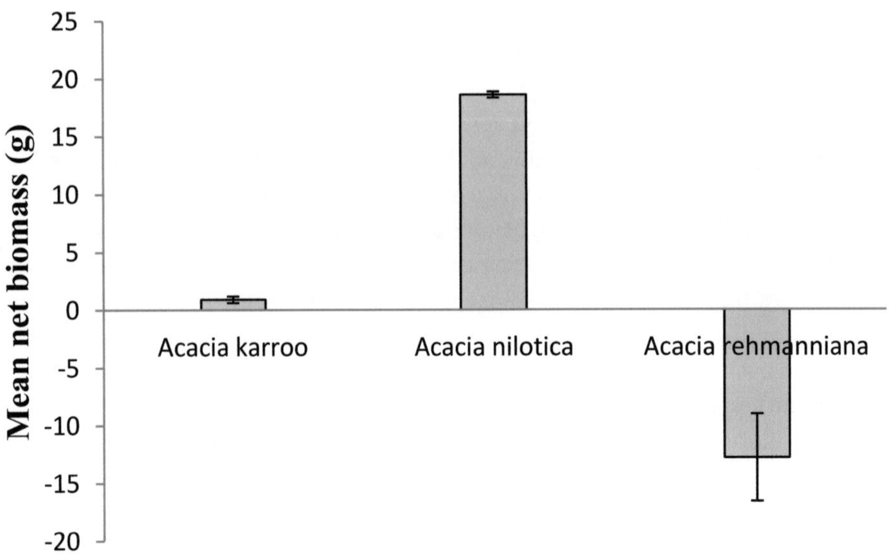

Figure 1. Mean (±SE) net biomass of *Acacia karoo*, *Acacia nilotica* and *Acacia rehmanniana* following shoot clipping in a semi-arid savanna. *Source*: Tsumele et al., 2009

Under-compensation of lost biomass may prevent further browsing while full- or over-compensation may increase forage availability and quality and thus initiate further browsing (Bowyer & Bowyer, 1997). Dube et al. (2009) also reported *A. nilotica* as more tolerant to herbivory than *Acacia nigrescens* and *Faidherbia albida*. Resprouting of shoots is intense following herbivory early in the growth season as plants have more time to recover before the end of the growth season. Shoot regrowth depends on the amount of carbohydrates that can be mobilized through photosynthesis or in carbohydrate reserves (Page & Whitham, 1987). Regrowth following defoliation is positively correlated with the carbohydrate status of the plant (Trlica & Singh, 1979) and resilience to defoliation depends on the rapidity with which reserves are restored (Dahl & Hyder, 1977). Compensatory regrowth of woody plants following herbivory occurs when nutrient and water resources are not limiting (Rosenthal & Kotanen, 1994). In the early growth season nutrient and water resources are abundant allowing plants to over-compensate lost biomass while late in the growth season the resources are inadequate leading to under-compensation (See Figure 2).

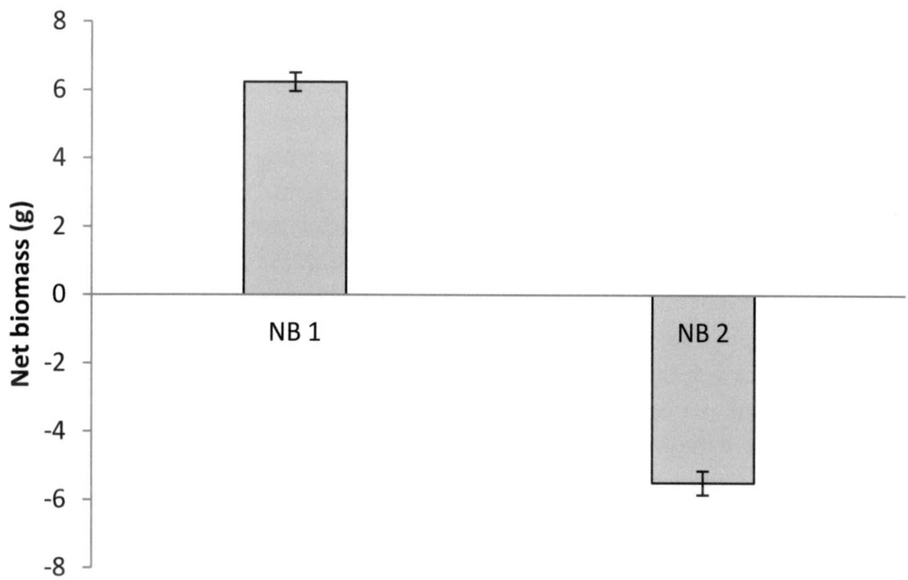

Figure 2. Mean (±SE) net biomass of *Grewia monticola* following shoot clipping during early growth season (NB 1) and late growth season (NB 2) in a semi-arid savanna. *Source:* Sebata et al., 2009

Herbivory during the early growth season coincides with nutrient flush enabling plants to benefit from energy mobilized from stored reserves. Scogings (2003) found defoliation during the early growth season to stimulate plant growth while growth of once-defoliated trees was not elevated above that of undefoliated trees when defoliation took place during the dormant season. Teague & Walker (1988) reported *Acacia karroo* as very sensitive to defoliation when carbohydrates reserves were at their lowest but very tolerant when reserves were high. Compensatory growth occurs when part of the photosynthetic material of a plant is removed resulting in more water and nutrients becoming available to the remaining photosynthetic components to increase their growth performance. Defoliation modifies the balance between growth promoting and inhibiting hormones in the plant resulting in various internal changes, such as increased photosynthesis, reduced rate of leaf senescence, changes in metabolite allocation and increased cell division and elongation which all contribute to increased growth (McNaughton, 1979; Teague & Walker, 1988). The growth following defoliation results in leaf replacement within weeks and then elevated levels of growth continue thereafter for more than a year, resulting in very large increases in leaf and shoot production compared to undefoliated plants (Teague & Walker, 1988). Compensatory regrowth following herbivory is considered an evolutionary response to herbivory (McNaughton, 1979). Crawley (1983) studied the implications of plant compensation at an ecosystem level and concluded that compensation following defoliation can improve ecosystem stability and increase the abundance of herbivores. Teague & Walker (1988) argued that compensation would benefit the plant only if herbivores fed on the plant intermittently or for a limited time each year, with the plants able to compensate once feeding has stopped. Strong plant compensatory growth cannot occur under continuous herbivore foraging.

Removal of the main shoots during browsing reduces apical dominance leading to the development of lateral shoots from activated dormant buds. Twig browsing in woody species can remove significant proportions of meristems resulting in fewer shoots in the following growth season (Bergstrom et al., 2000). The remaining shoots will experience less competition and thus grow larger and have higher nutrient concentrations than those on undamaged trees (Bergstrom et al., 2000; Rooke et al., 2004). Teague & Walker (1988) reported the increases in leaf and shoot of *A. karroo* plants following defoliation as due to the large increases of relatively few dominant shoots in the upper canopy. Danell et al. (1994) found leaf stripping of trees during the growth season to result in an increase in the number of shoots and a decrease in shoot size the following season. Some woody plants respond to severe defoliation from intensive browsing by producing many sprouts from basal shoots from the lower part of the stem enhancing the persistence of the plant (Bond & Midgley, 2001). This increases the plant's photosynthetic capability and creates the potential for increased juvenile recruitment. Resprouting shoots have been reported to have reduced defence compounds as a result of resources being allocated for fast growth at the expense of defence or the breakdown of existing defence compounds for use in growth (Coley et al., 1985).

Browsing reduces tree density, canopy cover and canopy diameter (Noumi et al., 2010) and affects tree regeneration (Mekuria et al., 1999). Fornara & du Toit (2008) reported *Acacia* trees at lightly browsed sites as having wider canopies and branches with longer internodes than trees at heavily browsed sites. The short internodes in the heavily browsed *Acacia* trees resulted in reduced canopy volume and increased side-branching on browsed shoots due to suppression of apical dominance (Du Toit et al., 1990). Browsing by megaherbivores e.g. African elephants (*Loxodonta africana*) reduces tree height resulting in a larger proportion of shoots and leaves becoming available within the browsing height of most terrestrial herbivores (Makhabu et al., 2006). Makhabu et al. (2006) reported impala (*Aepyceros melampus*) and greater kudu (*Tragelaphus strepsiceros*) as benefiting from the impacts of elephants in converting tall trees to short trees. More shoots were produced in heights reachable by both impala and kudu. Other studies have reported browsed trees as producing shoots with increased biomass per shoot (Bergström & Danell, 1987), increased nitrogen concentration and decreased concentration of secondary compounds like condensed tannins (Du Toit et al., 1990) compared with unbrowsed individuals. Eland at high densities prevent the recruitment of *Combretum apiculatum* from the 2.6 – 5.5 m height class to the >5.6 m height class (see Figure 3).

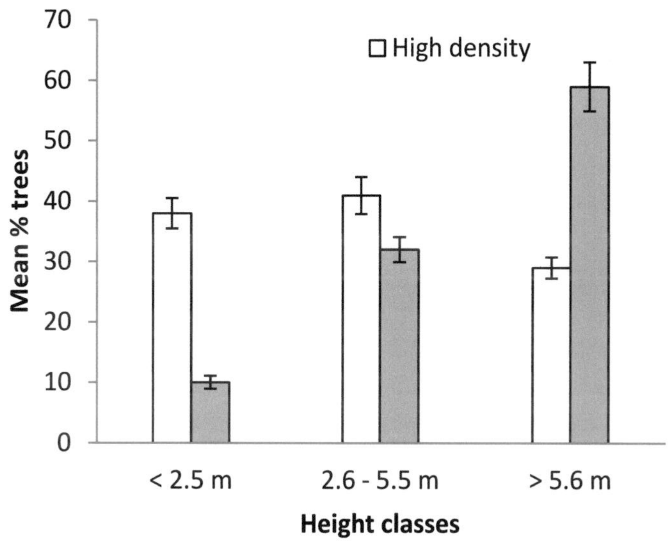

Figure 3. Mean (±SE) per cent of *Combretum apiculatum* trees in three height classes in relation to eland density. *Source:* Nyengera & Sebata 2009.

Shoot regrowths after defoliation have higher crude protein, phosphorus and biomass leading to repeated herbivory (Makhabu & Skarpe, 2006). Repeated browsing by megaherbivores such as the African elephant (*Loxodonta africana*) leads to the formation of low, intensely coppiced trees or stands of trees with high production of preferred browse (Makhabu et al., 2006). Rebrowsing means that the targeted trees suffer repeated damage and may eventually die or suffer reduced competitive ability relative to other woody plants (Skarpe & Hester, 2008). Heavy browsing by giraffe reduces tree growth rates increasing their susceptibility to drought (Birkett & Stevens-Wood, 2005). Fornara & du Toit (2008) reported high plant compensatory growth abilities of *Acacia nigrescens* as important for its persistence under heavy browsing in the Kruger National Park, South Africa. Herbivores also restrict the growth and the survival of young trees (Mwalyosi, 1990). Noumi et al. (2010) reported browsing as improving the regeneration of *Acacia tortilis* trees through the recruitment of new individuals. Skarpe (1990) argued that browsing accelerates tree growth in semi-arid savanna through reducing competition for moisture from herbaceous plants.

Herbivory may interfere with sexual reproduction in plants, either indirectly by changing physiology and allocation of resources, or directly by consumption of flower buds during the dormant season and flowers and fruits during the growth season (Skarpe & Hester, 2008; Fornara & du Toit, 2008). Herbivory results in plants allocating more resources to vegetative growth at the expense of sexual reproduction favouring species that reproduce vegetatively (Crawley, 1997). Goheen et al. (2007) reported herbivory as negatively affecting *Acacia drepanolobium* reproduction in an eastern African savanna.

3. Woody plant response to herbivory

Woody plants have evolved different strategies to reduce the negative effects of herbivory on their fitness (Rosenthal & Kotanen, 1994; Strauss & Agrawal, 1999). The strategies employed by plants to cope with herbivory can be classed into tolerance and avoidance mechanisms.

Tolerance strategies minimise the impacts of the damage (Hanley et al., 2007), with tolerant plants being generally palatable to the herbivores (Skarpe & Hester, 2008). Woody plants show tolerance to herbivory through morphological means such as quick replacement of lost leaves and shoots from protected meristems or through physiological processes such as compensatory photosynthesis and high and flexible rates of nutrient absorption (Hester et al., 2006). Re-sprouts have higher photosynthetic rates than older leaves. Teague (1989) reported *Acacia karroo* as relying on deep rooting, strong reserves and rapid growth to counter herbivory. Tolerance in plants is assumed to have little direct effect on herbivore fitness and is thus considered unlikely to trigger counter-adaptations in herbivores (Rosenthal & Kotanen, 1994).

Plants avoid being consumed by employing structural deterrents such as spines and thorns, biochemical compounds such as proanthocyanidins (condensed tannins) and internal constitutive defences such as lignin and cellulose, which also act as structural support. Lignin influences the physical toughness and digestibility of plants reducing intake rates (Jung & Allen, 1995; Scogings et al., 2004; Shipley & Spalinger, 1992). The structural deterrents are defined as spines when they are made of leaves and thorns when they are made of branches (Raven et al., 1999). Spines and thorns are the first line of defence against herbivores foraging on most woody plants in semi-arid savanna. They provide mechanical protection through injuring herbivores' mouths, digestive systems and other body parts. The presence of spines and thorns reduces the rate of herbivory by impeding stripping motions and forcing the herbivore to eat around the defence (Myers & Bazely, 1991; Wilson & Kerley, 2003a). Spinescent woody plants also have small leaves further reducing herbivore foraging efficiency since the reward received is seldom worth the time or energy needed to exploit it (Belovsky et al., 1991; Gowda, 1996). Plant spinescence increases with exposure to herbivory by large browsers as an induced defence (Milewski et al., 1991). Spines and thorns protect both leaves and axillary meristems (Gowda, 1996). Spine and thorn removal experiments have been carried out to demonstrate the protective value of these structures (Wilson & Kerley, 2003b; Hanley et al., 2007). Milewski et al. (1991) reported the removal of *Acacia drepanolobium* thorns as causing a threefold increase in mammalian browsing of new foliage. Increased rates of herbivory by bushbucks (*Tragelaphus scriptus*) and boergoats (*Capra hircus*) was also reported following the removal of thorns from spinescent shrub species in the Eastern Cape region of South Africa (Wilson & Kerley, 2003b).

Avoidance strategies also involve keeping most edible biomass beyond the reach of terrestrial herbivores. This means that the plant will have to survive herbivory before growing beyond the reach of the browsers. Woody plants growing in nutrient–rich environments are likely to grow above browsing height for most herbivores faster than trees in nutrient-poor environments, which will suffer browsing for a longer period (Danell et al., 1997). Woody plants growing in nutrient-poor environments have slow growth rates that limit their capacity to grow rapidly beyond the reach of most browsing mammals. They have developed strong defences for protection against herbivory (Coley et al., 1985; Teague, 1989; Borchard et al., 2011). Woody plants that grow in resource-rich environments often do not avoid herbivory, but develop tolerance traits to minimize the harmful effects of herbivory (Skarpe & Hester, 2008).

Storage of carbohydrates reserves in woody stems or underground is also a kind of escape strategy. Plants may also escape herbivory by association with either less palatable or more palatable species, depending on the foraging pattern of the herbivore (Hjalten et al., 1993; Hester et al., 2006). When palatable plants gain protection from their unpalatable neighbours the phenomenon is referred to as associational defence (McNaughton, 1978; Hjalten et al., 1993). However, palatable plants are usually susceptible to attack when they occur in a patch with unpalatable neighbours, a situation referred to as neighbour contrast susceptibility (Bergvall et al., 2006).

Plants do not respond passively to damage by herbivory. The optimal defence hypothesis predicts increases in defences in direct response to herbivory (Rhoades, 1979). Herbivore attack leads to decreased acceptability and plant nutritional quality (Malecheck & Provenza, 1983; Rhoades, 1985; Lundberg & Astrom, 1990). Plant defences will either reduce consumption rates or reduce the ability of herbivores to digest material once consumed (Belovsky et al., 1991; Robbins, 1993). Plants damaged by herbivores prevent further damage through an increase in digestion inhibiting compounds such condensed tannins (Cooper & Owen-Smith, 1985) and an increase in structural deterrents such as spines and thorns (Milewski et al., 1991). Some African woody species such as *A. karroo* have been shown to increase chemical defences following physical damage (Teague, 1989). Condensed tannins deter herbivory by giving plants an undesirable, astringent taste (Harborne, 1991; Bryant et al., 1992) or by reducing availability of protein and other nutrients (Robbins et al., 1987) through protecting plant cell walls from being degraded in the rumen of herbivores and inactivating digestive enzymes (Cooper & Owen-Smith, 1985). Milewski et al. (1991) reported branches of African *Acacia* trees that had been browsed by large herbivores as producing longer thorns and a greater density of thorns than inaccessible branches on the same trees. Rohner & Ward (1997) also reported intense herbivory of *Acacia tortilis* as increasing thorn length and density. Long thorns deter large herbivores by decreasing bite sizes and biting rates (Cooper & Owen-Smith, 1986; Belovsky et al., 1991; Gowda, 1996). Teague (1989) suggested that young shoots of *A. karroo* relied on chemical defences because their thorns were soft and offered little structural deterrence to the herbivores.

Both avoidance and tolerance involves costs for the plant such as in the building and maintenance of stores of energy and nutrients as well as of dormant buds that can be activated following herbivory (Bilbrough & Richards, 1993). Plant defences compete with growth and reproductive requirements for nitrogen and carbohydrate resources (Hanley et al., 2007). Owen-Smith & Cooper (1987) reported fewer plants as investing in both chemical and structural anti-herbivore defences to reduce costs to growth and reproduction. *Acacia tortilis* is heavily defended by both chemical and structural defences (Rohner & Ward, 1997; Sebata et al., 2011). Most *Acacia* species occur in areas of low fertility (Rohner & Ward, 1997) and adapt to these conditions by slow growth rates and efficient use of available nutrients (Coley et al., 1985), which may explain the ability of *A. tortilis* to invest in both types of defence.

4. Herbivore adaptations to plant defences

Herbivores need to develop ways of counteracting plant defences in order to utilise woody plants as browse (Hanley et al., 2007). Herbivores that forage on spinescent plants have smaller mouthparts to deal with the intricate task of removing small leaves from between dense assemblages of spines and thorns (Belovsky et al., 1991). Most browsing animals have agile lips and tongues that allow them to select leaves and avoid thorns (Gordon and Illius, 1988). For example goats with their mobile and narrow muzzle, can manoeuvre their mouths more easily among thorns to pluck small leaves, making thorns less effective in reducing cropping rates (Shipley et al., 1999; Cooper & Owen-Smith, 1986). Giraffe (*Giraffa camelopardalis)* foraging on spinescent *Acacia* trees is facilitated by the possession of a long flexible tongue (Hanley et al., 2007). Most ungulate herbivores in the semi-arid savanna where spinescence is most prevalent also have tough, leathery mouthparts, and nicitating eye membranes, both thought to be adaptations for coping with foraging on spinescent plants (Brown, 1960). Browsers foraging on spinescent plants may compensate for the reduced foraging efficiency by spending more time at plants of that species. Foraging on twig tips where growth is occurring and thorns are soft may also be adopted as a strategy to increase intake rates (Singer et al., 1994).

The evolution of a ruminant stomach can also be considered as an adaptation to plant defences since this allows the ungulates to digest fibrous plant material (Perez-Barberia et al., 2004). The ungulate stomach has symbiotic microorganisms and also releases cellulase enzymes which break down cellulose-rich cell wall fractions of plant material releasing volatile fatty acids that are immediately absorbed by the stomach (Hanley et al., 2007).

Some herbivores are able to develop behavioural and physiological counter adaptations against chemical plant defences (Iason & Villalba, 2006). For example, browsers such as goats secrete tannin-binding salivary proteins which counter the digestibility-reducing effect of ingested condensed tannins (Robbins et al., 1987). Tannin-binding salivary proteins contain a high proportion of proline, and proline-rich salivary proteins have a greater binding affinity for tannins than other proteins, and thus act to prevent tannins from interacting with other proteins in mammalian digestive systems (Shimada, 2006). The production of proline-rich proteins enhances cell wall (fiber) digestion of high-tannin forages by ungulates (Robbins et al., 1987).

5. Browse instantaneous intake rates

The foraging efficiency of browsers on different woody species can be defined in terms of the instantaneous intake rate (Wilson & Kerley, 2003b). Browse instantaneous intake rate is a product of bite size and bite rate and is influenced by plant characteristics. Different browse species will allow browsers to crop varying number and size of bites leading to highly variable instantaneous intake rates. Illius & Gordon (1990) estimated that browsers crop between 10 000 and 40 000 bites per day from different individual plants. Decisions made by the browser when selecting a bite have important consequences for its nutritional intake and hence fitness (Shipley et al., 1999). Most woody plants with nutritious forage have thorns or spines (Wilson & Kerley, 2003b). In semi-arid and arid African savannas thorny plants occur in areas with many large browsers (Grubb, 1992). Plant characteristics such as leaf size, thorn density and inter-thorn spacing (leaf accessibility) affect instantaneous intake rates through their effects on bite size and bite rate (See Table 1).

	r	r^2	Regression equation	
a) Intake rate *vs*				
Bite size	0.89	0.79	$y = 0.04 + 0.02x$	**
Bite rate	0.76	0.58	$y = 20.87 + 3.02x$	**
b) LAIN *vs*				
Bite rate	0.70	0.49	$y = 36.09 + 7.35x$	**
Bite size	0.45	0.20	$y = 0.14 + 0.03x$	**
Intake rate	0.62	0.38	$y = 4.97 + 1.63x$	**
c) Thorn density *vs*				
Bite size	- 0.57	0.33	$y = 0.12 - 0.03x$	*
Bite rate	- 0.66	0.43	$y = 31.79 - 6.42x$	*
Intake rate	- 0.69	0.48	$y = 3.86 - 1.70x$	**
d) Leaf size *vs*				
Intake rate	0.73	0.53	$y = 0.87 + 0.20x$	**
Bite size	0.60	0.36	$y = 0.05 + 0.05x$	*
Bite rate	0.51	0.26	$y = 22.2 + 7.13x$	ns

ns - not significant, * < 0.05, ** < 0.01. *Source:* Sebata & Ndlovu, 2010.

Table 1. Relationships (y = a + bx) of intake rates, leaf accessibility indices (LAIN), thorndensity and leaf size (y) of five woody species in a semi-arid southern African savanna and various browse intake rate parameters (x) achieved by goats when browsing on these plants (n = 14)

To achieve higher instantaneous intake rates browsers have to select browse species that allow large bite sizes and higher bite rates. Thus factors that constrain both bite size and bite rate will reduce instantaneous intake rates. Leaf accessibility and leaf size positively influenced bite size while thorn density had a negative effect (Table 1). Species with higher leaf accessibility allowed higher bite rates as the goats could easily maneuver their mouths between thorns when plucking the leaves. Thorns restricted goat muzzle movement slowing down the rate of browse harvesting (Belovsky et al., 1991). Thorns also force browsers to change foraging strategy from twig biting and leaf stripping to the less detrimental picking of leaves from between the thorns (Cooper & Owen-Smith, 1986; Gowda, 1996), reducing the loss of foliage to mammalian browsers. Browsers will achieve higher instantaneous intake rates through selecting species with higher leaf accessibility and larger leaves. However, handling time increases with increasing leaf size, suggesting that there is an optimum leaf size (Wilson & Kerley, 2003a).

6. Relationship between shoot morphology and herbivory

Shoot morphology has an influence on how plants protect themselves against loss of valuable nutrients and photosynthetic tissue to herbivores (Sebata & Ndlovu, 2012). Scogings et al. (2004) reported defences as being distributed among woody plants in semi-arid savannas according to shoot morphology because it affects the vulnerability of plant parts to browsers. Woody plants can be divided into two groups *viz.* those that produce all their new leaves on new long shoots (shoot-dominated species) and those that produce most of their new leaves in clusters on short shoots at the nodes of old unbrowsable branches (shoot-limited species) (Scogings et al., 2004). Shoot-dominated species depend on active apical buds to extend internodes and add new leaf area and should thus have higher concentrations of nutrients than shoot-limited species which simply add new leaf area without shoot elongation (Ganqa & Scogings, 2007; Scogings et al., 2004). Shoot-dominated species have more browseable shoots than shoot-limited species. Shoot-limited species tend to result in high bite rates and reduced instantaneous intake rates, while shoot-dominated species allow bigger bite sizes and relatively high instantaneous intake rates (Dziba et al., 2003). The apical meristems of shoot-dominated species are more vulnerable to herbivores than those of shoot-limited species (Dziba et al., 2003) and thus require better anti-herbivory defences (Rhoades, 1979). Plants and plant parts exposed to herbivores are expected to be better chemically defended than those protected by structural deterrents (Cooper & Owen-Smith, 1985). Goats have been shown to prefer shoot-limited over shoot-dominated species (See Figure 4).

The shoot-limited species are poorly defended chemically and depend on structural defences (thorns) which the goats are able to avoid using their mobile upper lips. Shoot-limited species have lower contents of plant secondary compounds (condensed tannins and fibre) and higher digestibility and rumen fermentation than shoot-limited species (Sebata & Ndlovu, 2012). Fibre enhances leaf toughness and reduces browsing (Jung & Allen, 1995; Shipley & Spalinger, 1992). Shoot-limited species also rapidly replace lost tissues through regrowth (Scogings et al., 2004). Shoot-limited and shoot-dominated species are able to adapt different anti-herbivory defences.

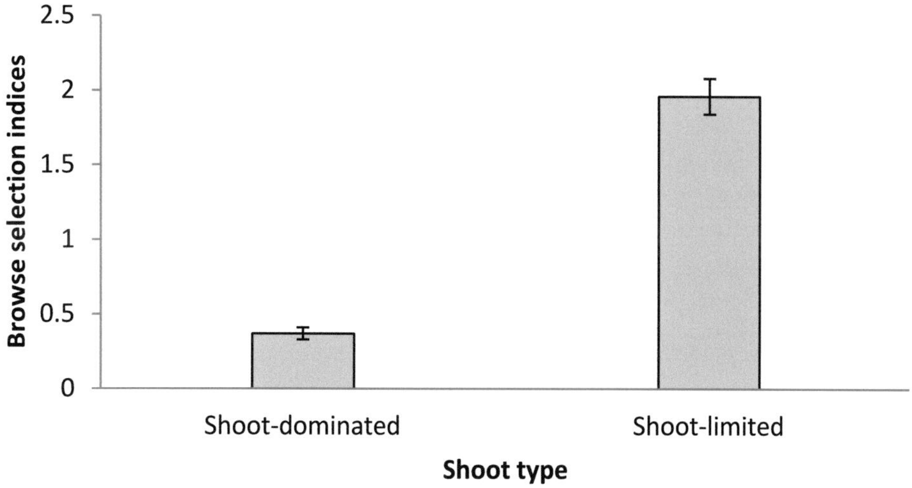

Figure 4. Browse selection indices of shoot-dominated and shoot-limited species in a semi-arid savanna ecosystem. *Source:* Sebata & Ndlovu 2012

7. Conclusion

Woody plants, at light stocking rates, are able to compensate biomass lost to herbivory. However, at high animal densities they may not be able to replace lost foliage, which could eventually lead to their mortality. Thus to maintain a positive herbivore-plant relationship ungulate populations in savanna ecosystems need to be regulated. Although herbivory stimulates woody plant resprouting, there is still need for defences against excessive defoliation. However, plant defences compete with growth needs requiring a balance in resource allocation. The allocation of nutrients and water resources to defence and growth is poorly understood necessitating further studies. The most effective herbivore adaptation to plant defences is selection of browse with low physical and chemical defences e.g. selecting shoot-limited over shoot-dominated woody species. The extent to which herbivore adaptations to plant defences allow ungulates to exploit the diverse woody plant resources needs to be studied. Woody plants in semi-arid savanna ecosystems are able to persist under intense herbivory due to key adaptations that include structural defences, chemical defences and compensatory growth abilities of the plants. The relationship between plant defences and high compensatory growth abilities of the plants are poorly understood. Structural defences are effective in limiting foliage loss to browsers and represent a cheap form of defence in semi-arid savannas.

Author details

Allan Sebata

Department of Forest Resources & Wildlife Management,
National University of Science & Technology (NUST), Ascot, Bulawayo, Zimbabwe

8. References

Belovsky, G.E.; Schmitz, O.J.; Slade, J.B. & Dawson, T.J. (1991). Effects of spines and thorns on Australian arid zone herbivores of different body masses. *Oecologia* 88: 521-528.

Belsky, A.J. (1986). Does herbivory benefit plants? A review of the evidence. *American Naturalist,* 127: 870–892.

Bergström, R. & Danell, K. (1987). Effects of simulated browsing by moose on morphology and biomass of two birch species. *Journal of Ecology* 75: 533–544.

Bergstrom, R.; Skarpe, C. & Danell, K. (2000). Plant responses and herbivory following simulated browsing and stem cutting of *Combretum apiculatum. Journal of Vegetation Science* 11: 409-414.

Bergvall, U.A.; Rautio, P.; Kesti, K.; Tuomi, J. & Leimar, O. (2006). Associational effects of plant defences in relation to within and between-patch food choice by a mammalian herbivore: neighbour contrast susceptibility and defence. *Oecologia* 147: 253–260.

Bilbrough, C.J. & Richards, J.H. (1993). Growth of sagebrush and bitterbrush following simulated winter browsing: mechanisms of tolerance. *Ecology* 74:481-492.

Birkett, A. & Stevens-Wood, B. (2005). Effects of low rainfall and browsing by large herbivores on an enclosed savannah habitat in Kenya. *African Journal of Ecology* 43:123–130.

Bond, W.J. & Midgley, J.J. (2001). Ecology of sprouting in woody plants: the persistence niche. *Trends in Ecological Evolution* 16: 45-51.

Borchard, F.; Berger, H.; Bunzel-Drüke, M. & Fartmann, T. (2011). Diversity of plant–animal interactions: Possibilities for a new plant defense indicator value? *Ecological Indicators* 11: 1311–1318.

Bowyer, J.W. & Bowyer, R.T. (1997). Effects of previous browsing on the selection of willow stems by Alaskan moose. *Alces* 33: 11–18.

Brown, G.D. (1960). Ants, acacias and browsing mammals. *Ecology* 41: 587–592.

Bryant, J.P.; Reichardt, P.B. & Clausen, T.P. (1992). Chemically mediated interactions between woody plants and browsing mammals. *Journal of Range Management* 45:18–24.

Coley, P.D.; Bryant, J.P. & Chapin, F.S. (1985). Resource availability and plant antiherbivore defence. *Science* 230: 895-899.

Cooper, S.M. & Owen-Smith, N. (1985). Condensed tannins deter feeding by browsing ruminants in a South-African savanna. *Oecologia* 67: 142–146.

Cooper, S.M. & Owen-Smith, N. (1986). Effects of plant spinescence on large mammalian herbivores. *Oecologia* 68: 446-455.

Cornell, H.V. & Hawkins, B.A. (2003). Herbivore responses to plant secondary compounds: a test of phytochemical coevolution theory. *American Naturalist* 161:507–522.

Crawley, M.J. (1983). *Herbivory: the dynamics of animal-plant interactions.* Blackwell Scientific publications, Oxford.

Crawley, M.J. (1997). *Plant ecology.* Blackwell Scientific publications, Oxford.

Dahl, B.E. & Hyder, D.N. (1977). *Developmental morphology and management implications.* In: Sosebee, R.E. (Ed.), Rangeland plant physiology. Society for Range Management. Denver, Colorado.

Danell, K.; Bergstrom, R. & Edenius, L. (1994). Effects of large mammalian browsers on architecture, biomass, and nutrients of woody plants. *Journal of Mammalogy* 75: 833-844.

Danell, K.; Haukioja, E. & Huss-Danell, K. (1997). Morphological and chemical responses of mountain birch leaves and shoots to winter browsing along a gradient of plant productivity. *Ecoscience* 4: 296-303.

Dangerfield, J.M. & Modukanele, B. (1996). Overcompensation by *Acacia erubescens* in response to simulated browsing. *Journal of Tropical Ecology* 12: 905–908.

Dube, S.; Mlambo, D. &Sebata, A. (2009). Response of *Faidherbia albida* (Del.) A. Chev., *Acacia nigrescens* Oliver. and *Acacia nilotica* (L.) Willd ex Del. seedlings to simulated cotyledon and shoot herbivory in a semi-arid savanna in Zimbabwe. *African Journal of Ecology* 48: 361–367.

Du Toit, J.T. (2003). *Large herbivores and savanna heterogeneity*. In: Du Toit, J.T., Rogers, K.H., Biggs, H.C. (Eds.), The Kruger Experience: Ecology and Management of Savanna Heterogeneity. Island Press, Washington, DC, pp. 292–309.

Du Toit, J.T.; Bryant, J.P. & Frisby, K. (1990). Regrowth and palatability of *Acacia* shoots following pruning by African Savanna browsers. *Ecology* 71: 140–154.

Dziba, L.E.; Scogings, P.F.; Gordon, I.J. & Raats, J.G. (2003). Effects of season and breed on browse species intake rates and diet selection by goats in the False Thornveld of the Eastern Cape, South Africa. *Small Ruminant Research* 47: 17–30.

Fornara, D.A. & du Toit, J.T. (2008). Community-level interactions between ungulate browsers and woody plants in an African savanna dominated by palatable-spinescent *Acacia* trees. *Journal of Arid Environments* 72: 534–545.

Ganqa, N.M. & Scogings, P.F. (2007). Forage quality, twig diameter, and growth habit of woody plants browsed by black rhinoceros in semi-arid sub-tropical thicket, South Africa. *Journal of Arid Environment* 70: 514–526.

Goheen, J.R.; Young, T.P.; Keesing, F. & Palmer, T.M. (2007). Consequences of herbivory by native ungulates for the reproduction of a savanna tree. *Journal of Ecology* 95:129–138.

Gordon, I.J. & Illius, A.W. (1988). Incisor arcade structure and diet selection in ruminants. *Functional Ecology* 2: 15–22.

Gowda, J.H., 1996. Spines of *Acacia tortilis*: what do they defend and how? *Oikos* 77: 279–284.

Grubb, P.J. (1992). A positive distrust in simplicity – lessons from plant defences and from competition among plants and among animals. *Journal of Ecology* 80: 585–610.

Hanley, M.E.; Lamont, B.B.; Fairbanks, M.M. & Rafferty, C.M. (2007). Plant structural traits and their role in anti-herbivore defence. *Perspectives in Plant Ecology, Evolution and Systematics* 8: 157–178.

Harborne, J.B. (1991). *The chemical basis of plant defence*. In: Palo, R.T., Robbins, C.T. (Eds.), Plant Defences against Mammalian Herbivory. CRC Press, Boca Raton, FL, pp. 192.

Hartley, S.E. & Jones, C.G. (1997). *Plant chemistry and herbivory, or why is the World green?* In: Crawley, M.J. (Ed.), Plant Ecology. Blackwell, Oxford, pp. 284–324.

Hester, A.J.; Bergman, M.; Iason, G.R. & Moen, R. (2006). *Impacts of large herbivores on plant community structure and dynamics.* In: Danell, K., Bergstrom, R., Duncan, P., Pastor, J. (eds) Large herbivore ecology and ecosystem dynamics. Cambridge University Press, Cambridge, pp 97-141.

Hjalten, J., Danell, K. & Lundberg, P. (1993). Herbivore avoidance by association – vole and hare utilization of woody plants. *Oikos* 68: 125-131.

Iason, G.R. & Villalba, J.J. (2006). Behavioural strategies of mammal herbivores against plant secondary metabolities: the avoidance-tolerance continuum. *Journal of Chemical Ecology* 32: 1115-1132.

Illius, A.W. & Gordon, I.J. (1990). *Constraints on diet selection and foraging behavior in mammalian herbivores.* In: Hughes, R.E. (Ed.), Behavioural Mechanisms of Food Selection. Springer, Berlin, Germany, pp. 369-393.

Jung, H.G. & Allen, M.S. (1995). Characteristics of plant cell walls affecting intake and digestibility of forages by ruminants. *Journal of Animal Science* 73: 2774–2790.

Lundberg, P. & Astrom, M. (1990). Low nutritive quality as a defence against optimally foraging herbivores. *American Naturalist* 135: 547–561.

Makhabu, S.W. & Skarpe, C. (2006). Rebrowsing by elephants three years after simulated browsing on five woody plant species in northern Botswana. *South African Journal of Wildlife Research* 36:99-102.

Makhabu, S.W; Skarpe, C. & Hytteborn, H. (2006). Elephant impact on shoot distribution on trees and on rebrowsing by smaller browsers. *Acta Oecologia.* 30:136-146.

Malecheck, J.C. & Provenza, F.D. (1983). Feeding behaviour and nutrition of goats on rangelands. *World Annual Review* 3: 38-48.

Maschinski, J. & Whitham, T.G. (1989). The continuum of plant responses to herbivory: the influence of plant association, nutrient availability, and timing. *American Naturalist,* 134:1–19.

McNaughton, S.J. (1978). Serengeti ungulates: feeding selectivity influences the effectiveness of plant defence guilds. *Science* 199: 806–807.

McNaughton, S.J. (1979). Grazing as an optimization process: grass–ungulate relationships in the Serengeti. *American Naturalist,* 113: 691–703.

Mekuria, A.; Demel, T. & Mats, O. (1999). Soil seed flora, germination and regeneration pattern of woody species in *Acacia* woodland of the Rift Valley in Ethiopia. *Journal of Arid Environments* 43: 411–435.

Milewski, A.V.; Young, T.P. & Madden, D. (1991). Thorns as induced defences: experimental evidence. *Oecologia* 86: 70-75.

Mwalyosi, R.B.B. (1990). The dynamic ecology of *Acacia tortilis* woodland in Lake Manyara National Park, Tanzania. *African Journal of Ecology* 28: 189–199.

Myers, J.H. & Bazeley, D. (1991). *Thorns, spines, prickles and hairs: are they stimulated by herbivory and do they deter herbivores?* In: Tallamyr, D.J., Raup, M.J. (eds) Phytochemical induction by herbivores. Academic Press, New York, pp 326-343.

Noumi, Z.; Touzard, B.; Michalet, R. & Chaieb, M. (2010). The effects of browsing on the structure of *Acacia tortilis* (Forssk.) Hayne ssp. *raddiana* (Savi) Brenan along a gradient of water availability in arid zones of Tunisia. *Journal of Arid Environments* 74, 625–631.

Nyengera, R. & Sebata, A. (2009). Effect of eland density and foraging on *Combretum apiculatum* physiognomy in a semi-arid savannah. *African Journal of Ecology,* 48:45–50

Owen-Smith, N. & Cooper, S.M. (1987). Palatability of woody plants to browsing ruminants in a South African savanna. *Ecology* 68: 319–331.

Page, K.N. & Whitham, T.G. (1987). Overcompensation in response to mammalian herbivory: the advantage of being eaten. *American Naturalist* 129: 407–416.

Perez-Barberia, F.J.; Elston, D.A.; Gordon, I.J. & Illius, A.W. (2004). The evolution of phylogenetic differences in the efficiency of digestion in ruminants. *Proceedings of the Royal Society. London Series* B 271: 1081–1090.

Raven, P.H.; Evert, R.F. & Eichhorn, S.E. (1999). *Biology of Plants*, sixth ed. W. H. Freeman and Company, New York.

Rhoades, D.F. (1979). *Evolution of plant chemical defence against herbivores*. In: Rosenthal, G.A., Janzen, D.H. (Eds.), Herbivores: Their Interaction with Secondary Plant Metabolites. Academic Press, Orlando, pp. 3–54.

Rhoades, D.F. (1985). Offensive-defensive interactions between herbivores and plants: Their relevance in herbivore population dynamics and ecological theory. *American Naturalist* 125: 205-238.

Robbins, C.T. (1993). *Wildlife Feeding and Nutrition*. Academic Press, San Diego.

Robbins, C.T.; Mole, S.; Hagerman, A.E. & Hanley, T.A. (1987). Role of tannins in defending plants against ruminants: reduction in dry matter digestion. *Ecology* 68: 1606–1615.

Rohner, C. & Ward, D. (1997). Chemical and mechanical defence against herbivory in two sympatric species of desert Acacia. *Journal of Vegetation Science* 8: 717–726.

Rooke, T.; Bergstrom, R.; Skarpe, C. & Danell, K. (2004). Morphological responses of woody species to simulated twig-browsing in Botswana. *Journal of Tropical Ecology* 20: 281-289.

Rosenthal, G.A. & Janzen, D.H. (1979). *Herbivores: Their Interaction with Secondary Plant Metabolites.* Academic Press, New York.

Rosenthal, J.P. & Kotanen, P.M. (1994). Terrestrial plant tolerance to herbivory. *Trends in Ecological Evolution* 9: 145-148.

Scholes, R.J. (1997). *Savanna.* In: Cowling, R.M., Richardson, D.M., Pierce, S.M. (Eds.), Vegetation of Southern Africa. Cambridge University Press, Cambridge, pp. 258–277.

Scogings, P.F. (2003). *Impacts of ruminants on woody plants in African savannas: an overview.* In: Allsopp, N., Palmer, A.R., Milton, S.J., Kirkman, K.P., Kerley, G.I.H., Hurt, C.R., Brown, C.J. (Eds.), Rangelands in the New Millenium. VII International Rangeland Congress, Durban, South Africa, pp. 955–957.

Scogings, P.F. & Mopipi, K. (2008). Effects of water, grass and N on responses of *Acacia karroo* seedlings to early wet season simulated browsing: Aboveground growth and biomass allocation. *Journal of Arid Environments* 72:509–522.

Scogings, P.F.; Dziba, L.E. & Gordon, I.J. (2004). Leaf chemistry of woody plants in relation to season, canopy retention and goat browsing in a semiarid subtropical savanna. *Austral Ecology* 29: 278–286.

Sebata, A. & Ndlovu, L.R. (2010). Effect of leaf size, thorn density and leaf accessibility on instantaneous intake rates of five woody species browsed by Matebele goats (*Capra hircus* L) in a semi-arid savanna, Zimbabwe. *Journal of Arid Environment* 74: 1281-1286.

Sebata, A. & Ndlovu, L.R. (2012). Effect of shoot morphology on browse selection by free ranging goats in a semi-arid savanna. *Livestock Science* 144: 96–102.

Sebata, A.; Nyathi, P. & Mlambo, D. (2009). Growth responses of *Grewia flavescens* Juss. (Sandpaper Raisin) and *Grewia monticola* Sond. (Grey Grewia) (Tiliaceae) to shoot clipping in a semi-arid Southern African savanna. *African Journal of Ecology* 47: 794–796.

Sebata, A.; Ndlovu, L.R. & Dube, J.S. (2011). Chemical composition, in vitro dry matter digestibility and in vitro gas production of five woody species browsed by Matebele goats (*Capra hircus* L.) in a semi-arid savanna, Zimbabwe. *Animal Feed Science and Technology* 170: 122– 125.

Shimada, T. (2006). Salivary proteins as a defence against dietary tannins. *Journal of Chemical Ecology* 32:1149–1163.

Shipley, L.A. & Spalinger, D.E. (1992). Mechanisms of browsing in dense food patches: effects of plant and animal morphology on intake rate. *Canadian Journal of Zoology* 70:1743–1752.

Shipley, L.A.; Illius, A.W.; Danell, K.; Hobbs, N.T. & Spalinger, D.E. (1999). Predicting bite size selection of mammalian herbivores: a test of a general model of diet optimization. *Oikos* 84: 55-68.

Singer, F.J.; Mark, L.C. & Cates, R.C. (1994). Ungulate herbivory of willows on Yellowstone's northern winter range. *Journal of Range Management* 47: 435–443.

Skarpe, C. (1990). Shrub layer dynamics under different behaviors density in arid savannas, Botswana. *Journal of Applied Ecology* 27: 873–885.

Skarpe, C. (1992). Dynamics of savanna ecosystems. *Journal of Vegetation Science* 3:293–300.

Skarpe, C. & Hester, A. (2008). *Plant traits, browsing and grazing herbivores, and vegetation dynamics.* In: Gordon, I.J., Prins, H.H.T. (eds) The ecology of browsing and grazing. Springer-Verlag Berlin Heidelberg. pp 217-261.

Strauss, S.Y. & Agrawal, A.A. (1999). The ecology and evolution of plant tolerance to herbivory. *Trends in Ecological Evolution* 14: 179-185.

Teague, W.R. (1989). Patterns of selection of *Acacia karroo* by goats and changes in tannin levels and *in vitro* digestibility following defoliation. *Journal of the Grassland Society of Southern Africa* 6: 230-235.

Teague, W.R. & Walker, B.H. (1988). Effect of intensity of defoliation by goats at different phenophases on leaf and shoot growth of *Acacia karroo* Hayne. *Journal of the Grassland Society of Southern Africa* 5: 197-206.

Trlica, M.J. & Singh, J.S. (1979). *Translocation of assimilates and creation, distribution and utilization of reserves.* In: Goodall, D.W., Perry, R.A. (Eds.), Arid land ecosystems: structure, functioning and management. Cambridge University Press.

Tsumele, J., Mlambo, D., Sebata, A. (2009). Responses of three *Acacia* species to simulated herbivory in a semi-arid southern African savanna. *African Journal of Ecology* 45: 324–326.

Walker, B.H. (1985). *Structure and function of savannas: an overview.* In: Tothill, J.C., Mott, J.J. (Eds.), Ecology and Management of the World's Savannas. Australian Academy of Science and CAB, Farnham Royal, Canberra, pp. 83–91.

Wilson, S.L. & Kerley, G.I.H. (2003a). Bite diameter selection by thicket browsers: the effect of body size and plant morphology on forage intake and quality. *Forest Ecology and Management* 181: 51–65.

Wilson, S.L., Kerley, G.I.H. (2003b). The effect of plant spinescence on the foraging efficiency of bushbuck and boergoats: browsers of similar body size. *Journal of Arid Environments* 55: 150-158.

Chapter 2

Are the Species of the Genus *Avicennia L.* (Acanthaceae) a "Superhost" Plants of Gall-Inducing Arthropods in Mangrove Forests?

Rita de Cassia Oliveira dos Santos, Marcus Emanuel Barroncas Fernandes and Marlucia Bonifácio Martins

Additional information is available at the end of the chapter

http://dx.doi.org/10.5772/55454

1. Introduction

Some plant species, especially angiosperms, present infestation by invertebrates that modify the general appearance of their vegetative and/or reproductive parts, known as galls. The galls are the result of abnormal growth of cells, tissues or organs due to an increase in cell volume (hypertrophy) and/or in the number of cells (hyperplasia) in response to feeding or to stimuli caused by foreign bodies, except for other inducing agents such as fungi and bacteria, which lead to amorphous tumor formation (Rohfritsch & Shorthouse 1982; Dreger-Jauffret & Shorthouse 1992; Raman *et al.* 2005; Raman 2007).

Gall-inducing arthropods have a highly specific relationship with their host plants, and normally attack only a single or a few closely-related plant species (Dreger-Jauffret & Shorthouse 1992). This degree of specialization facilitates the recognition of the diversity of gall-inducing insects, for example, in a given locality or plant species (Carneiro *et al.* 2009). The presence in some plant communities of species that support a relative rich fauna of gall-inducing insects has resulted in these plants being referred to as "superhost", whose local and regional distribution have a decisive influence on the local and regional galling diversity (Veldtman & McGeoch, 2003; Espírito-Santo *et al.*, 2007; Mendonça, 2007).

Some plant species, from a range of biogeographic regions and ecosystems, present a wide variety of gall morphotypes (Fernandes & Price 1988; Waring & Price 1989; Blanche 2000; Stone *et al.* 2002; Cuevas-Reyes *et al.* 2003). In the Neotropical region, for example, "superhosts" have been identified in a number of different habitat types, such as highland rocky grassland ("campo rupestre"), involving the species *Baccharis concinna* (Carneiro *et al.* 2005) and *Baccharis pseudomyriocefala* (Lara *et al.* 2002), and *Copaifera langsdorffii* (Oliveira *et al.*

2008), associated with litholic habitats ("canga"), as well as halophytic formations, where examples include *Eugenia umbelliflora* (Maia 2001b; Monteiro *et al.* 2004) and *Guapira opposita* (Oliveira & Maia 2005). In the temperate zone, Waring & Price (1989) and Gagné & Waring (1990) have also identified the creosote bush (*Larrea tridentata*) as a "superhost" for Cecidomyiidae (Diptera) species of the *Asphondylia auripila* group.

The vast majority of the studies of herbivory by insects in mangrove forests have focused on leaf-chewing species (Cannicci *et al.* 2008), adding the fact that little is known about interactions involving endophytic forms such as leaf miners or gall-inducing species (Gonçalves-Alvim *et al.* 2001; Burrows 2003; Menezes & Peixoto 2009).

The genus *Avicennia* has pantropical distribution, comprising the following species: *A. alba* Blume, *A. bicolor* Standley, *A. eucalyptifolia* (Zipp. ex Miq.) Moldenke, *A. germinans* (L.) Stearn, *A. integra* Duke, *A. lanata* Ridley, *A. marina* (Forks.) Vierh., *A. officinalis* L., *A. rumphiana* Hallier f. e *A. schaueriana* Stapf and Leechman ex Moldenke (WORMS 2010). Based on the fact that the genus *Avicennia* presents a great variety of gall morphotypes and hence many gall-inducing arthropods, the present study aims to review this plant-gall association describing what is known so far, and to verify whether the species of *Avicennia* are "superhost" plants of tropical mangrove forests.

2. Gall-inducing arthropods and their hosts in mangrove worldwide

Of the nine species of gall-inducing arthropods described for mangrove plants in different parts of the world, two species (Acari: Eriophyidae) were described in the genus *Laguncularia* L. (Combretaceae) (Flechtmann *et al.* 2007) and seven species (Diptera: Cecidomyiidae) in the genus *Avicennia* L. (Acanthaceae) (Cook 1909; Felt 1921; Gagné & Law 1998; Gagné & Etienne 1996). Besides, there are twenty-two morphotyped already described, totalizing 29 species of Cecidomyiidae recorded on four species of *Avicennia* (Table 1). Table 1 also presents three other groups of arthropods (Acari, Hemiptera, and Hymenoptera), which have been registered associated with mangroves around the world, and the family Cecidomyiidae stands out as the main group of galling in this ecosystem.

In fact, the species of the genus *Avicennia* have been identified as hosts of gall- inducing cecidomyiids in the Neotropical region since the beginning of the twentieth century. Gagné & Etienne (1996) reviewed the data on this phenomenon and proposed that the insect *Cecidomyia avicenniae* (Diptera, Cecidomyiidae), described by Cook in 1909 in *Avicennia nitida* in Cuba and Central America, should be reassigned to the species *Meunieriella avicenniae* (Cook). They also emphasized the fact that Tomlinson (1986) considered *A. nitida* to be a synonym of *A. germinans*.

Galls similar to those described by Cook at the beginning of the century were also identified subsequently in *A. tomentosa* in the Brazilian state of Bahia (Tavares 1918), and in *Avicennia officinalis* in French Guiana (Houard 1924). However, the two mangrove species identified in these studies were in fact *A. germinans* or *Avicennia schaueriana*, given that *A. tomentosa* is a synonym of *A. germinans* and *A. officinalis* is found only in the Old World (Tomlinson 1986).

Plant host	Inducing taxon	Reference
A. germinans	CECIDOMYIIDAE	
(6 spp.)	*Meunieriella avicenniae*	Tavares 1918; Houard 1924; Gagné & Etienne 1996
	Undet. sp. A Undet. sp. B	Jiménez 2004
	Undet. sp. B	Jiménez 2004
	Undet. sp. C	Jiménez 2004
	PSYLLIDAE	
	Telmapsylla minuta	Jiménez 2004
	ACARI	
	Undet. sp. A	Jiménez 2004
A. marina	CECIDOMYIIDAE	
(16 spp.)	*Actilasioptera coronata*	Gagné & Law 1998
	Actilasioptera falcaria	Kathiresan 2003
	Actilasioptera pustulata	Gagné & Law 1998
	Actilasioptera subfolium	Gagné & Law 1998
	Actilasioptera tuberculata	Gagné & Law 1998
	Actilasioptera tumidifolium	Gagné & Law 1998
	Undet. sp. A Undet. sp. B Undet. sp. C Undet. sp. D Undet. sp. E	Sharma & Das 1994; Sharma *et al.* 2003
	Undet. sp. B	Sharma & Das 1994; Sharma *et al.* 2003
	Undet. sp. C	Sharma & Das 1994; Sharma *et al.* 2003
	Undet. sp. D	Sharma & Das 1994; Sharma *et al.* 2003
	Undet. sp. E	Sharma & Das 1994; Sharma *et al.* 2003
	Undet. sp. F	Sharma & Das 1994; Sharma *et al.* 2003
	ERYOPHYIDAE	
	Undet. sp. A	Gagné & Law 1998
	COCCIDAE	
	Undet. sp. A	Gagné & Law 1998
	ACARI or INSECTA	
	Undet. sp. A	Gagné & Law 1998
	HYMENOPTERA	

Plant host	Inducing taxon	Reference
	Undet. sp. A	Sharma & Das 1994; Sharma *et al.* 2003
A. officinalis	CECIDOMYIIDAE	
(4 spp.)	*Actilasioptera falcaria*	Raw & Murphy 1990
	Undet. sp. A	Raw & Murphy 1990
	Undet. sp. B	Raw & Murphy 1990
	ERYOPHYIDAE	
	Eriophyes sp.	Raw & Murphy 1990
A. schaueriana	Undet. sp. A	Maia *et al.* 2008
(3 spp.)	*Meunierilla aviceniae*	Menezes & Peixoto 2009
	Undet. sp. B	Present study

Table 1. Gall-inducing arthropods on species of *Avicennia*

In addition, Gagné (1994) identified rounded and smooth galls in *A. germinans* in Florida (USA) and Belize, while Gagné & Etienne (1996) identified the same gall morphotypes in specimens of *A. germinans* in Guadalupe and San Martin, in Central America. In Belize, Central America, Farnsworth & Ellison (1991) identified two gall morphotypes in *A. germinans,* but without identifying the inducing agent. Jiménez (2004) highlighted the presence of leaf galls induced by *Telmapsylla minuta* (Psyllidae: Homoptera), by three undetermined species of Cecidomyiidae, and an unidentified mite species in *A. germinans,* in Costa Rica.

In Ranong, Thailand, the occurrence of four gall-inducing insects in *A. officinalis* was recorded, being two induced by undetermined species of Cecidomyiidae, one by an unidentified mite species belonging to the family Eriophyidae and another by *Stefaniella falcaria* Felt (Diptera: Cecidomyiidae) (Rau & Murphy, 1990). Subsequently, Gagné & Law (1998) reviewed a material from Java and redescribed and transferred the species *Stefaniella falcaria* to the new genus *Actilasioptera.*

Regarding *A. schaueriana,* at least three species of Cecidomyiidae are involved in leaf gall-inducing in this species, two in southeastern Brazil, *Meunieriella avicennae* (= *Cecidomyia avicennae*) and an undetermined (Maia *et al.*, 2008; Menezes & Peixoto, 2009), and another in the northern region (present study).

Gagné & Law (1998) described from Queensland, Australia, five gall-inducing insects of the family Cecidomyiidae in *A. marina,* all of which belong to a new genus, *Actiolasioptera* (*A. coronata* Gagné, 1998; *A. pustulata* Gagné, 1998; *A. subfolium* Gagné, 1998; *A. tuberculata* Gagné, 1998; *A. tumidifolium* Gagné, 1998). In addition to these midge species, a eriophyid mite (Acari), a coccid (Homoptera), and an unknown arthropod gall were identified in *A. marina* in Australia. In addition, Sharma & Das (1984) and Sharma *et al.* (2003) recorded six undetermined species of Cecidomyiidae (leaf gall) and an unidentified species of

Hymenoptera (stem gall) for *A. marina*, from the coastal region of Andaman and Vikhroli, Marahashtra, India. Kathiresan (2003) also recorded *Stephaniella falcaria* (=*Actilasioptera falcaria*) in the mangrove forests of Pichavaram, southeastern coast of India. Burrows (2003), investigating herbivory in mangroves near Townsville, Queensland, Australia, recorded at least ten gall morphotypes to this mangrove species, without identifying, however, the gall-inducing agents.

3. Gall morphotypes in *Avicennia*

The vast majority of gall-inducing arthropods is restricted to a single host plant species, thus corroborating the idea that the gall morphotype can be used as reliable substitutes of gall-inducing species. In addition, polymorphism of galls, which could lead to the occurrence of failures in the identification of galls, appears to be a rather rare phenomenon (Carneiro *et al.* 2009).

Numerous surveys have been conducted in different regions of Brazil in an attempt to identify the diversity of gall-inducing agents in different ecosystems (Fernandes & Price 1988; Gonçalves-Alvim & Fernandes 2001; Fernandes & Negreiros 2006; Araújo *et al.* 2007; Maia *et al.* 2008; Carneiro *et al.* 2009). Given that some species of the genus *Avicennia* support many gall-inducing arthropods throughout their geographic distribution, it is expected a greater diversity of gall-inducing arthropods and a wide variety of gall morphotypes growing on their organs (Table 2).

4. New records of gall morphotypes in *Avicennia* of the north coast Brazil

Given the wide distribution of the genus *Avicennia* around the world, it is clear that the interaction between galls and *Avicennia* species is an import gap in our understanding of the role of these trees in the mangrove system. At this topic, in addition to the literature review, new records of gall morphotypes found in *Avicennia* on the Ajuruteua Peninsula (00°57,9'S-46°44,2'W), on Brazil's Amazon coast (Fig. 1) will be also presented. Data collection was carried out in October/2010 and leaves from 20 individuals of *A. germinans* were collected simultaneously.

A total of 11,448 leaves were counted on the 20 specimens of *A. germinans*. Of this total, only 17% (n=1,970) had galls, which were classified in 14 distinct morphotypes, based on their coloration and morphological features (Fig. 2 and Table 3).

Among 4,787 morphotyped galls, conical ones were the most common (n=1,101 – 23%), while disc galls were the rarest (n=9 – 0.2%). The number of galls per leaf varied from one to 41, with an average of 4.7±1.5 galls. Table 3 also indicates that an additional 2,313 galls – almost a third of the total of 7,100 recorded in the study – were not identified, due to be either damaged (DAM) or in an initial stage of development (ISD). The total number of galls in the individuals of *A. germinans* sampled at the Furo do Taici must be higher, although stem galls have been observed in different branches, they were not counted in the present study.

Plant host	Organ	Form	Geographic location	Reference
A. germinans or *A. schaueriana*	Leaf	Round and smooth on the upper surface and warty on the lower surface, a craterlike exit hole eventually develops	Cuba	Cook 1909; Tavares 1918; Houard, 1924
A. germinans	Leaf	No description	Belize	Farnsworth & Ellison 1991
A. germinans	Leaf	No description	Costa Rica	Jiménez 2004
A. germinans	Leaf	No description	Costa Rica	Jiménez 2004
A. germinans	Leaf	No description	Costa Rica	Jiménez 2004
A. officinalis	Leaf	Gregarious small gall	Thailand	Rau & Murphy 1990
A. officinalis	Leaf	No description	India	Katherisan 2003
A. officinalis	Leaf	Small globular gall	Thailand	Rau & Murphy 1990
A. officinalis	Leaf, Petiole and/or shoot	Large globular gall	Thailand	Rau & Murphy 1990
A. officinalis	Leaf (abaxial surface of midvein)	Keel-like gall	Thailand	Rau & Murphy 1990
A. officinalis	Leaf	Large flattened gall	Thailand	Rau & Murphy 1990
A. officinalis	Leaf	Small, widely scattered pouch gall	Thailand	Rau & Murphy 1990
A. officinalis	Leaf	The gall is a 1 cm diameter swelling situated on or very near leaf midvein, is unevenly round and apparent both leaf surfaces	Java	Gagné & Law 1998
A. germinans	Leaf	Round and smooth on the upper surface and warty on the lower surface, a craterlike exit hole eventually develops	Belize and USA	Gagné 1994
A. germinans	Leaf	Round and smooth on the upper surface and warty on the lower surface, a craterlike exit hole eventually develops	Guadeloupe and Saint Martin	Gagné & Etienne 1996

Plant host	Organ	Form	Geographic location	Reference
A. germinans	Leaf	Spheroid on the abaxial and adaxial surfaces	Brazil	Gonçalves-Alvim *et al.* 2001
A. marina	Leaf	No description	India	Katherisan 2003
A. marina	Leaf	Small, unevenly hemispheroid and warty on the upper surface, craterlike on the lower surface	Australia	Gagné & Law 1998
A. marina	Leaf	Circular and flat on the upper surface, nearly evenly hemispherical on the lower surface	Australia	Gagné & Law 1998
A. marina	Leaf	Large, mostly soft, simple leaf swelling, apparent on only the lower surface	Australia	Gagné & Law 1998
A. marina	Leaf	Circular basally, with one or more elongate, conical projections arising from it on the upper surface	Australia	Gagné & Law 1998
A. marina	Leaf	Large, simple, convex leaf swelling, apparent on both leaf surfaces	Australia	Gagné & Law 1998; Burrows 2003
A. marina	Leaf	No description	Australia	Gagné & Law 1998
A. marina	Leaf	No description	Australia	Gagné & Law 1998
A. marina	Leaf	No description	Australia	Gagné & Law 1998
A. marina	Leaf	Edge gall	Australia	Burrows 2003
A. marina	Leaf	Tower/Spike gall	Australia	Burrows 2003
A. marina	Leaf	Cabbage gall	Australia	Burrows 2003
A. marina	Leaf	Yellow lamp gall	Australia	Burrows 2003
A. marina	Leaf	Marble gall	Australia	Burrows 2003
A. marina	Leaf	Midvein gall	Australia	Burrows 2003
A. marina	Leaf	Acne gall	Australia	Burrows 2003
A. marina	Leaf	Raised-pit gall	Australia	Burrows 2003
A. marina	Leaf	Pimple gall	Australia	Burrows 2003
A. marina	Stem	Stem gall	Australia	Burrows 2003
A. schaueriana	Leaf	Unilocular globoid gall	Brazil	Maia *et al.* 2008;
A. schaueriana	Leaf	No description	Brazil	Menezes & Peixoto 2009
A. schaueriana	Leaf	No description	Brazil	Present study

Table 2. Gall morphotypes of *Avicennia*

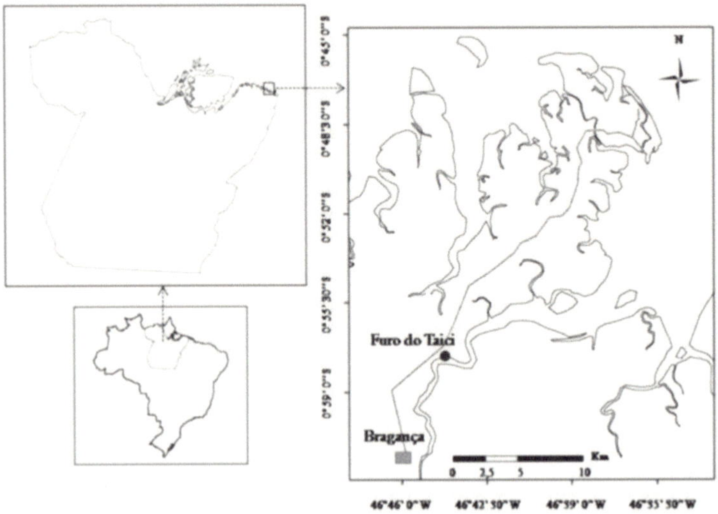

Figure 1. Location of the study area (Furo do Taici) on the Ajuruteua Peninsula, Bragança, Pará.

Figure 2. 2-17. Gall morphotypes observed on the leaves of *A. germinans*: GAB: Globular - 2) adaxial surface, 3) abaxial surface; GBD: Globular - 4) adaxial surface, 5) abaxial surface; GCC: Globular with a central concavity - 6) adaxial surface, 7) abaxial surface; GAD: Globular - 8) adaxial surface, 9) abaxial surface; GLO: Globoid - 10) adaxial surface, 11) abaxial surface; GTA: Globular with a tapering free portion - 12) adaxial surface, 13) abaxial surface; FLA: Flattened - 14) adaxial surface, 15) abaxial surface; CYL: Cylindrical - 16) adaxial surface, 17) abaxial surface.

Figure 3. 18-29. Gall morphotypes observed on the leaves of *A. germinans*: CON: Conical - 18) adaxial surface, 19) abaxial surface; VLC: Volcanic crater - 20) adaxial superfície, 21) abaxial surface; DIS: Discoid - 22) adaxial surface, 23) abaxial surface; MDR: Midrib - 24) adaxial surface, 25) abaxial surface; PET: Petiolar - 26: adaxial surface, 27) abaxial surface; AOV: Aggregated Ovoid Gall - 28) adaxial surface, 29) abaxial surface.

Six of the 14 morphotypes identified during the study were globular in shape (to varying degrees): i) lobular gall on the abaxial surface of the leaf – monolocular, with opening on the adaxial surface (GAB); ii) globular gall on the abaxial and adaxial surfaces – globular on both surfaces, with opening on the adaxial surface (GBD); iii) globular gall with central concavity – monolocular, with opening on the adaxial surface (GCC); iv) globular gall on the adaxial surface of the leaf – monolocular (GAD); v) globoid gall – monolocular, normally very close to one another and more globular on the adaxial surface (GLO); vi) globular gall – monolocular, globular base on the abaxial surface with a tapering free portion (GTA); vii) flattened gall with a slit-like opening – monolocular, globoid on the adaxial surface, with opening on the adaxial surface (FLA); viii) cylindrical gall – globoid on the adaxial surface, with opening at the base of the cylindrical portion, on the abaxial surface (CYL); ix) conical gall – monolocular gall with a conical shape on both surfaces of the leaf and opening on the adaxial surface, coloration changes from greenish to purplish during senescence (CON); x) volcanic crater gall – globoid on the adaxial surface, opening in the tubular portion on the abaxial surface, often grouped (VLC); xi) discoid gall – monolocular, globoid only on the adaxial surface (DIS); xii) gall on the midrib of the leaf – appears as a thickening of the midrib, often individually but also in closely-spaced agglomerations, opening on the adaxial surface (MDR); xiii) petiolar gall – appears as a thickening of the petiole (PET); xiv) aggregated ovoid gall – distributed very close together in circular, flower-shaped groups (AOV) (Table 4).

Finally, a subsample of the fourteen morphotyped galls were separated and placed in plastic pots to await the emergence of gall-inducing arthropods. Of the gall-inducing insects isolated from the leaves of *A. germinans*, seven morphotypes were identified as belonging to the family Cecidomyiidae (Table 3).

	Morphotype	**Number of galls**	**Mean galls per plant (±SD; n=20)**	**%**
Identified	CON	1,101	55.1±42.8	15.5
	AOV	922	46.1±43.9	13.0
	GAB	685	34.3±32.4	9.6
	GCC	591	29.6±28.5	8.3
	MDR	398	19.9±17.2	5.6
	GLO	397	19.9±34.3	5.6
	PET	237	11.9±29.7	3.3
	GBD	146	7.3±9.9	2.1
	CYL	141	7.1±16.8	2.0
	VLC	55	2.8±4.3	0.8
	GAD	54	2.7±4.9	0.8
	GTA	36	1.8±2.5	0.5
	FLA	15	0.8±2.0	0.2
	DIS	9	0.5±0.6	0.1
Subtotal		4,787		
Unidentified	ISD	1,837	91.9±71.9	25.8
	DAM	476	23.8±23.5	6.7
Subtotal		2,313		
Total		7,100	355.4±244.6	100

Table 3. Number (total, mean ± standard deviation) and relative frequency (%) of galls by morphotype found on 1,970 leaves of *Avicennia germinans* (L.) Stearn (Acanthaceae), in the Furo do Taici of the Ajuruteua Peninsula in Bragança, in the Brazilian state of Pará. IDENTIFIED: CON = Conical gall; AOV = Aggregated Ovoid gall; GAB = Globular gall – Abaxial surface; GCC = Globular gall with a Central Concavity; MDR = Midrib gall; GLO = Globoid gall; PET = Petiolar gall; GBD = Globular gall – Abaxial and Adaxial surfaces; CYL = Cylindrical gall; VLC = Volcanic Crater gall; GAD = Globular gall – Adaxial surface; GTA = Globular gall with a tapering free portion; FLA = Flattened gall; DIS = Discoid gall. UNIDENTIFIED: ISD = Initial stage of development; DAM = Damaged; SD = Standard Deviation.

Morphotype	shape	Organ/Location	Color	Pusbescence	Chamber	Ocurrence
CON	Conical	Leaf	Green/Purple	Glabrous	One	Isolated
AOV	Aggregated ovoid	Leaf	Green	Glabrous	One	Coalescent
GAB	Globular concavity	Leaf	Green	Glabrous	One	Isolated
GCC	Globular with central concavity	Leaf	Green	Glabrous	One	Isolated
MDR	Midrib	Leaf	Green	Glabrous	Several	Isolated
GLO	Globoid	Leaf	Green	Glabrous	One	Coalescent
PET	Petiolar	Leaf/Petiole	Green	Glabrous	Several	Isolated
GBD	Globular	Leaf`	Green	Glabrous	One	Isolated
CYL	Cylindrical	Leaf	Green	Glabrous	One	Isolated
VLC	Volcanic crater	Leaf	Green	Glabrous	One	Isolated/ Coalescent
GAD	Globular woody	Leaf	Green	Glabrous	One	Isolated
GTA	Globular with tapering free portion	Leaf	Green	Glabrous	One	Isolated
FLA	Flattened	Leaf	Green	Glabrous	One	Isolated
DIS	Discoid	Leaf	Green	Glabrous	One	Isolated

Table 4. Description of the gall morphotypes identified on the leaves of *Avicennia germinans* (L.) Stearn (Acanthaceae), in the Furo do Taici of the Ajuruteua Peninsula in Bragança, in the Brazilian state of Pará. CON = Conical gall; AOV = Aggregated Ovoid gall; GAB = Globular gall – Abaxial surface; GCC = Globular gall with a Central Concavity; MDR = Midrib gall; GLO = Globoid gall; PET = Petiolar gall;.GBD = Globular gall – Abaxial and Adaxial surfaces; CYL: Cylindrical gall; VLC = Volcanic Crater gall; GAD = Globular gall – Adaxial surface; GTA = Globular gall with a tapering free portion; FLA = Flattened gall; DIS = Discoid gall;

5. Conclusions

The genus *Avicennia* presents a pioneer group of species which is highly tolerant of salinity (Hogarth 1999), and has leaves with high levels of total nitrogen (Medina *et al.* 2001), low levels of secondary compounds (Roth 1992), and high leaf productivity with less energy investment (Cannicci *et al.* 2008). In addition to the wide distribution of this genus, where *Avicennia* species occur they are often abundant and the dominant species. These characteristics, together with the reduced plant diversity of the mangrove ecosystem on a regional scale (Menezes *et al.* 2008), are probably among the key factors to determine the preference of endophytic herbivores for this species.

However, Blanche (2000) notes that the available studies have reported different effects of plant species richness on the diversity of gall-inducing insects, and according to Veldtman & McGeoch (2003), in some areas taxonomic composition of the vegetation appears to be more important than species richness.

At present, of the ten currently recognized species of *Avicennia*, four have already been registered with galls: *A. germinans, A. marina, A. officinalis,* and *A. schaueriana.* In total, 44 gall morphotypes have already been recorded for species of *Avicennia* (Table 2), and therefore it must be considered as a "superhost" genus. The terminology "superhost" for a botanical genus has been previously proposed by Mendonça (2007).

Avicennia germinans and *A. marina* are, by far, the mangrove species with the greatest known variety of gall-inducing arthropods, with 22 and 19 galls, respectively, which doubtless characterizes both species as "superhost" plants. In the case of *A. germinans,* the categories DAM and ISD, together with the stem galls, suggest that this particular species may have an even larger number of gall morphotypes. In addition, *A. officinalis* and *A. schaueriana,* may also be considered potential "superhost" plants, since available records showed four and three species of gall-inducing arthropods associated with both species, respectively (Table 1 and 2). The species of *Avicennia* are similar with respect to their chemical, morphological, anatomical and ecological traits, which favor its infestation by several species of galling arthropods in different geographic regions (Tomlinson, 1986; Burrows, 2003). This fact becomes even more pronounced in areas that have low mangrove plant diversity and where other plant species have characteristics that prevent colonization by arthropods (e.g. sclerophyllous leaves and high amounts of secondary compounds), as in the genus *Rhizophora.*

Thus, it is important to bear in mind that *Avicennia* species appears to have a similar role in the trophic chain of the endophytic herbivores of the mangrove forest. Of the ten species of the genus *Avicennia,* only four have been recorded on the literature. Thus, it is likely that with the increasing progress the work on the interaction of arthropods with this botanical genus, it also increases the number of records of the endophytic herbivores.

The effect of arthropod herbivore activities may negative and positively impact both the mangrove trees and the ecosystem. Cannicci *et al.* (2008) pointed out that herbivory is usually considered to be a negative impact, due to the fact that they are more apparent and readily measured than positive ones. Regarding the fact that many gall-inducing organisms are associated with the genus *Avicennia,* it may be a considerable positive contribution to the overall diversity of herbivores in mangroves. Likewise, the premature abscission of a large quantity of leaf material (Burrows 2003) and the conversion of leaves into frass by caterpillars (Fernandes *et al.* 2009) may cause positive impacts on either individual or ecosystem, respectively, by providing high leaf yield for the trees and rich nutrient supply for the mangrove *per se*.

Author details

Rita de Cassia Oliveira dos Santos, Marcus Emanuel Barroncas Fernandes and Marlucia Bonifácio Martins
Universidade Federal do Pará, Brazil
Museu Paraense Emílio Goeldi, Brazil

Acknowledgement

We thank Fundação de Amparo à Pesquisa do Estado do Pará (FAPESPA) (Edital Nº 004/2007) and Instituto Internacional de Estudos Brasileiros (IIEB) – Programa BECA (Scholarship – B/2007/02/BDP/05), for the financial support to RCOS.

6. References

Araújo, W. S., Gomes-Klein, V. L. & Santos, B. B. (2007). Galhas entomógenas associadas à vegetação do parque estadual da serra dos pireneus, Pirenópolis, Goiás, Brasil. *Revista Brasileira de Biociências,* 5: 45-47.

Blanche, K. R. (2000). Diversity of insect-induced galls along a temperature-rainfall gradient in the tropical savannah region of the Northern Territory, Australia. *Austral Ecology,* 25: 311–318.

Burrows, D. W. (2003). *The role of insect leaf herbivory on the mangroves Avicennia marina and Rhizophora stylosa.* Ph.D. thesis, James Cook University, 238 p.

Cannicci, S.; Burrows, D; Fratini, S; Smith III, T. J.; Offenberg, J. & Dahdouh-Guebas, F. (2008). Faunal impact on vegetation structure and ecosystem function in mangrove forests: A review. *Aquatic Botany,* 89: 186–200.

Carneiro, M. A. A.; Fernandes, G. W. & De Souza, O. F. F. (2005). Convergence in the Variation of Local and Regional Galling Species Richness. *Neotropical Entomology,* 34:547–553.

Carneiro, M. A. A.; Branco, C. S. A.; Braga, C. E. D.; Almada, E. D.; Costa, M. B. M.; Maia, V. C. & Fernandes, G. W. (2009). Gall midge species (Diptera,Cecidomyiidae) host-plant specialists? *Revista Brasileira de Entomologia,* 53: 365–378.

Cook, M. T. (1909). Some insects gall of Cuba. *Estación Central Agronomica de Cuba,* 2: 143–146.

Cuevas-Reyes, P.; Siebe, C.; Martinez-Ramos, M. & Oyama, K. (2003). Species richness of gall-forming insects in a tropical rain forest: correlations with plant diversity and soil fertility. *Biodiversity and Conservation,* 12: 411–422.

Dreger-Jauffret, F. & Shorthouse, J. D. (1992). Diversity of gall-inducing insects and their galls, p. 8-33. *In*: J. D. Shorthouse & O. Rohfritsch (Eds.). *Biology of insect-induced galls,* New York, Oxford University Press, xi+285 p.

Espírito-Santo, M. M.; Neves, F. S.; Andrade-Neto, F. R. & Fernandes G. W. (2007). Plant architecture and meristem dynamics as the mechanisms determining the diversity of gall-inducing insects. *Oecologia,* 153: 353–364.

Farnsworth, E. J. & Ellison, A. M. (1991). Patterns of herbivory in Belizean mangrove swamps. *Biotropica,* 23: 555–567.

Felt, E.E. 1921. Javanese gall midges. *Treubia,* 1: 139-151.

Fernandes, G. W. & Price, P. W. (1988). Biogeographical gradients in galling species richness: test of hypotheses. *Oecologia* (Berl.), 76: 161–167.

Fernandes, G. W. & Negreiros, D. (2006). A comunidade de insetos galhadores da RPPN fazenda bulcão, Aimorés, Minas Gerais, Brasil. *Lundiana,* 7:111-120.

Fernandes, M. E. B.; Nascimento, A. A. M. & Carvalho, M. L. (2009). Effects of herbivory by *Hyblaea puera* (Hyblaeidea: Lepidoptera) on litter production in the mangrove on coast of Brazilian Amazonia. *Journal of Tropical Ecology,* 25: 337–339.

Flechtmann, C.H.W., Santos-Mendonça, I.V. & Almeida-cortez, J.S. (2007). A new species of Brachendus (Acari, Eriophyidae) associated with the white mangrove, Laguncularia racemosa (Combretaceae), in Brazil. *International Journal Acarology* 33:195-198.

Gagné, R. J. (1994). *The gall midges of neotropical region*. Ithaca, Comstock, xiv+352 p.

Gagné, R. J. & Etienne, J. (1996). *Meunieriella avicenniae* (Cook) (Diptera: Cecidomyiidae) the leaf gall marker of black mangrove in the American tropics. *Proceedings Entomological Society of Washington,* 98: 527–532.

Gagné, R. J. & Law L. J. (1998). *Actilasioptera* (Diptera: Cecidomyiidae), a new genus for Australiasian and Asian gall midges of grey mangroves, *Avicennia* spp. (Avicenniaceae), p. 22–35 *In*: Csoka, G., Mattson, W.J., Stone, G.N. and Price, P.W (eds.). *The Biology of Gall-Inducing Arthropods*. Minnesota, U.S. Dept. of Agriculture, Forest Service, North Central Research Station. 329 p.

Gagné, R. J. & Waring, G. L. (1990). The *Asphondylia* (Cecidomyiidae: Diptera) of creosote bush (*Larrea tridentata*) in North America. *Proceedings Entomological Society Washington,* 92: 649–671.

Gonçalves-Alvim, S. J. & Fernandes, G. W. (2001). Comunidades de insetos galhadores (Insecta) em diferentes fisionomias do cerrado em Minas Gerais, Brasil. *Revista Brasileira de Zoologia,* 18 (supl. 1): 289-305.

Gonçalves-Alvim, S. J.; Santos, M. C. F. V. & Fernandes, G. W. (2001). Leaf gall abundance on *Avicennia germinans* (Avicenniaceae) along interstitial salinity gradient. *Biotropica,* 33: 69–77.

Hogarth, P. J. (1999). *The biology of mangroves*. New York, Oxford University Press, 288 p.

Houard, C. (1924). Les collections cécidologiques du laboratoire d'entomologie du Muséum d'Histoire Naturelle de Paris: galles de la Guyane Française. *Marcellia,* 21:97–128.

Jiménez, J. A. (2004). Mangrove forests under dry seasonal climates in Costa Rica. In: Frankie, G. W., Mata, A. & Vinson, S. B. (Eds.). *Biodiversity Conservation in Costa Rica: Learning the lessons in a seasonal dry Forest.* Berkely and Los Angeles, California. University of California Press. 343 p.

Kathiresan, K. (2003). Insect folivory in mangroves. *Indian Journal of Marine Sciences,* 32: 237-239.

Lara, A. C. F.; Fernandes, G. W. & Gonçalves-Alvim, S. J. (2002). Tests of hypotheses on patterns of gall distribution along an altitudinal gradient. *Tropical Zoology,* 15: 219–232.

Maia, V. C. (2001b). The gall midges (Diptera, Cecidomyiidae) from three restingas of Rio de Janeiro State, Brazil. *Revista Brasileira de Zoologia,* 8: 583–630.

Maia, V. C.; Magenta, M. A. G. & Martins, S. E.. (2008). Ocorrência e caracterização de galhas de insetos em áreas de restinga de Bertioga (São Paulo, Brasil). *Biota Neotropica,* 8: 167–197.

Medina, E.; Giarrizo, T.; Menezes, M.; Carvalho Lira, M.; Carvalho, E. A.; Peres, A.; Silva, B. A.; Vilhena, R.; Reise, A. & Braga, F. C. (2001). Mangal communities of the "Salgado Paraense": Ecological heterogeneity along the Bragança peninsula assessed through soil and leaf analyses. *Amazoniana,* 16: 397–416.

Mendonça Jr., M. S. (2007). Plant diversity and galling arthropod diversity searching for taxonomic patterns in an animal-plant interaction in the Neotropics. Bol. Soc. Argent. Bot. 42: 3-4.

Menezes, M. P. M.; Berger, U. & Mehlig, U. (2008). Mangrove vegetation in Amazonia: a review of studies from the coast of Pará and Maranhão States, north Brazil. *Acta Amazonica,* 38: 403–420.

Menezes, L. F. T. & Peixoto, A. L.. (2009). Leaf damage in a mangrove swamp at Sepetiba Bay, Rio de Janeiro, Brazil. *Revista Brasileira Botânica,* 32: 715–724.

Monteiro, R. F.; Oda, R. A.; Constantino, P. A. L. & Narahara, K. L.. (2004). Galhas: diversidade, especificidade e distribuição, p. 127–141. *In*: Rocha, C. F. D.; Esteves, F. A. & Scarano, F. R. (Orgs.). *Pesquisas de longa Duração na Restinga de Jurubatiba: Ecologia, História Natural e Conservação,* v.1., p. 127-141. São Carlos, RiMa Editora, vii+376 p.

Oliveira, D. C.; Drummond, M. M.; Moreira, A. S. F. P.; Soares, G. L. G.& Isaias, R. M. S. (2008). Potencialidades morfogênicas de *Copaifera langsdorffii* Desf. (Fabaceae): super-hospedeira de herbívoros galhadores. *Revista Biologia Neotropical,* 5: 31–39.

Oliveira, J. C. & Maia, V. C. (2005). Ocorrência e caracterização de galhas de insetos na restinga de Grumari (Rio de Janeiro, RJ, Brasil). *Arquivos do Museu Nacional,* 63: 669–675.

Raman, A. (2007). Insect-induced plant galls of India: unresolved questions. *Current Science,* 92: 748–757.

Raman, A.; Schaefer, C. W. & Withers, T. M. (2005). *Biology, ecology, and evolution of gall-inducing arthropods,* Vol. 1 and 2. New Hampshire, Science Publishers Inc., xxi+817 p.

Rau, M. T. & Murphy, D. H. (1990). Herbivore attack on mangrove plants at Ranong. *Mangrove Ecosystems Occasional Papers,* No. 7: 25–36.

Rohfritsch, O. & Shorthouse, J. D. (1982). Insect galls, p. 131–152. *In:* G. Kahl & J. S. chell (eds.). *Molecular biology of plant tumors,* New York, Academic Press, xxiv+615 p.

Roth, I. 1992. Leaf structure: coastal vegetation and mangroves of Venezuela. Berlin, Gebrüder Borntraeger, 172 p.

Sharma, R. M. & Das, A. K. (1984). Further contribution to the knowledge of Zoocecidia of mangrove *Avicennia marina* (Forsk.) Vier. *Records of the Zoological Survey of India,* 81: 123–126.

Sharma, R. M., Joshi, P. V. & Shindikar, M. (2003). First report on plant galls (zoocecidia) from mangrove swamps of Vikhroli, Maharashtra. *Zoos' Print Journal,* 18: 1217–1219.

Stone, G. N.; Schönrogge, K.; Atkinson, R. J.; Bellido, D. & Pujade-Villar, J. (2002). The population biology of oak gall wasps (Hymenoptera: Cynipidae). *Annual Review of Entomology*, 47: 633–668.

Tavares, J. S. (1918). Cecidologia brazileira. Cecidias que se criam nas plantas das famílias das Verbenaceae, Euphorbiaceae, Malvaceae, Anacardiaceae, Labiatae, Rosaceae, Anonaceae, Ampelidaceae, Bignoniaceae, Aristolochiaceae e Solanaceae. *Broteria: Série Zoológica*, 16: 21–68.

Tomlinson, P. B. (1986). *The botany of mangroves*. Cambridge, Cambridge University Press, 413p.

Veldtman, R. & McGeoch, M. A. (2003). Gall forming insect species richness along a non-scleromorphic vegetation rainfall gradient in South Africa: The importance of plant community composition. Australian Ecology 28: 1-13.

Waring, G. L. & Price, P. W. (1989). Parasitoid pressure and the radiation of a gall-forming group (Cecidomyiidae: *Asphondylia* spp.) on creosote bush (*Larrea tridentata*). *Oecologia*, 79: 293–299.

WORMS. (2010). World Mangroves database. In: Dahdouh-Guebas F. (Ed.) (2010). Accessed through: World Register of Marine Species 32P32p://www.marinespecies.org/aphia.php?p=taxdetails&id=235131 on 2011-02-11.

Chapter 3

Are Ephippid Fish a "Sleeping Functional Group"? – Herbivory Habits by Four Ephippidae Species Based on Stomach Contents Analysis

Breno Barros, Yoichi Sakai, Hiroaki Hashimoto, Kenji Gushima, Yrlan Oliveira, Fernando Araújo Abrunhosa and Marcelo Vallinoto

Additional information is available at the end of the chapter

http://dx.doi.org/10.5772/47848

1. Introduction

Ephippidae fish are commonly classified as being omnivorous, though tending to carnivore habits fishes (Burgess, 1978; Heemstra, 2001; Kuiter & Debelius, 2001). The group is broadly distributed in sub-tropical and tropical coastal regions, comprising eight genera and 16 species (Nelson, 2006), where the larger genera are *Platax*, with five species described all from the Indo-Pacific (Kishimoto et al., 1988; Nakabo, 2002), and *Chaetodipterus*, with three species: one occurring in the eastern Pacific; and two reported from the Atlantic Ocean (Burgess, 1978).

The literature concerning feeding habits and feeding behavior of ephippid fish is still very scarce. Recent studies, however, show that the trophic classification of the group is controversial. Depending on the ontogenetic stage, and on specific environmental conditions, juvenile *Platax orbicularis*, were observed to switch from mainly herbivorous habits, during daylight, to carnivorous habits at night (Barros et al., 2008; 2011). Other studies based solely on stomach contents have identified mostly plant material in the stomachs of individuals at the same growth stage (Nanjo et al., 2008). Zooplankton consisted in the main food item identified in stomachs belonging to juveniles of another *Platax* species, *P. boersii* (Nanake et al., 2011). The latter is generally classified as carnivores, feeding mainly on benthic prey. All five *Platax* species are known from the Indo-Pacific, yet recent studies have reported sporadic occurrences in the Mediterranean Sea, indicating the group invasive potential (Bilecenoglu & Kaya, 2006; Golani et al., 2011).

Misleading information regarding diet and feeding habits of *Chaetodipterus* species are also reported in the available literature. While *C. faber* sampled from the coast of South Caroline, in the US, were observed to feed mainly on hydroids (Hayse, 1990), those from north-eastern Brazil showed preference for a more herbivorous, algae-rich diet (Bittencourt, 1980; Couto & Vaconcelos Filho, 1980), yet none of these reports exclude other food sources in their results.

Moreover, Bellwood et al. (2006) have suggested ephippid fish as belonging to a "sleeping functional group", where individual fish have the potential to explore algae-rich substrates for food, such as phase-shifted corals, helping with the recovering process of the latter environments, via removing the thick algae layer from whitened corals, as the observed in a few adult individuals of *P. pinnatus* at the Great Reef Barrier, Australia. The authors have suggested the entire Family Ephippidae might play such a role in damaged coral reef environments, including the genus *Chaetodipterus* in Atlantic coastal waters.

The present review aims to verify if ephippid fish should be classified as a potential functional group, examining both the available literature on Ephippid fish, as well as original data on the stomach contents of juveniles and adults of five ephippid species from four locations in Japanese and Brazilian coastal waters. Feeding plasticity, feeding behavior and the Group potential on playing a functional role on these coastal environments are discussed.

2. Feeding plasticity and trophic classification of Ephippid fish

Despite being a relatively small group, information on feeding habits, feeding behavior and diet of the Ephippidae is quite limited, being available mostly from technical reports based on trawl samples or bycatch material. As both methods are often limited in sample number, the resulting literature is then sparse and confusing: While there is plenty of information on trophic habits of the two main genera, *Platax* and *Chaetodipterus*, there are a number of incongruences on such reports. Furthermore, detailed information regarding the remaining genera is rare, virtually absent in the literature. In Table 1 we summarize the state of the art concerning the available knowledge on the trophic classification of the Ephippidae.

Establishing a general classification for an entire fish group is always controversial, as feeding habits may rely on several biological aspects of a given species, as ontogenetic stage, habitat conditions during settlement, etc., according to both morphological and environmental constraints for particular sizes (Gerking, 1994; Diana, 1995; Russo et al., 2008). Even for the closely related group Acanthuridae (Holcroft & Wiley, 2008), sister-group of the Ephippidae, and generally known as an herbivore group, a few species have a mixed diet, based on both zooplankton and algae (Choat et al., 2004).

According to the literature, a general classification for the trophic habits of ephippid fish is a difficult, almost impossible task. This is especially due to the contrasting information on some taxa, indicating a very plastic diet, which may include both animal and plant-based food sources, as well as different behavioral strategies, even in supposedly specialized species, such as those of the genus *Platax* (Bellwood et al., 2006; Barros et al., 2008, 2011). Detritivorous habits by *P. boersii* and *R. pentanemus* are also strong evidences supporting feeding plasticity: late juveniles and adults of *P. boersii* are commonly observed to chase the green turtle *Chelonia mydas* to feed on their algae-rich faeces, (B. Barros, per. obs.); and *R. pentanemus* were reported to feed on sewage material (Robins et al., 1991).

Taxon	Occurrence	Trophic classification	References
Chaetodipterus faber		Omnivore*	Bittencourt (1980) Couto & Vasconcelos Filho (1980) Hayse (1990)
Chaetodipterus lippei	Eastern Atlantic	-	-
Chaetodipterus zonatus	Eastern Pacific	Carnivore	Schneider (1995) de La Cruz Agüero et al. (1997)
Ephippus goreensis	Eastern Atlantic	Carnivore	Allen (1981) Desoutter (1990)
Ephippus orbis	Indo-West Pacific	Carnivore	Masuda et al. (1984) Maugé (1984) Kuronuma & Abe (1988) Lieske & Myers (1994)
Parasepttus panamensis	Eastern Pacific	-	-
Platax batavianus	Indo-West Pacific	-	-
Platax boersii	Indo-West Pacific	Omnivore**	Kuiter & Debelius (2001)
Platax orbicularis	Indo-West Pacific	Omnivore*	Myers (1991) Kuiter & Debelius (2001) Barros et al. (2008, 2011)
Platax pinnatus	Indo-West Pacific	Carnivore*	Kuiter & Debelius (2001) Bellwood et al. (2006)
Platax teira	Indo-West Pacific	Carnivore	Myers (1991)
Proteracanthus sarissophorus	Western-Central Pacific	-	-
Rhinoprenes pentanemus	Western-Central Pacific	Herbivore**	Robins et al. (1991)
Tripterodon orbis	Western Indian	Carnivore	Fischer et al. (1990)
Zabidius novemaculeatus	Indo-Pacific	-	-

Table 1. List of all 15 Ephippidae species (following Nelson, 2006), with their respective trophic classification, according to the available literature. Single asterisks indicate dubious literature regarding trophic habits, suggesting both herbivorous and carnivorous habits by these species. Double asterisks indicate detritivorous habits as well. Hyphens indicate no available data on diet or feeding habits.

Morphological attributes of the cranial anatomy of *Platax* and *Chaetodipterus*, are similar to those of Family Scaridae, with very specialized swelling of ethmoid, frontals and supraoccipital bones, shortening of the lower jaw, and short dentary and articular bones, which provide great biting power (Gregory, 1933). Browsing activities over algae turfs by juvenile *P. orbicularis* were also observed to feed on algae similar to the feeding manner of the Scaridae (Barros et al., 2008).

3. Stomach contents and herbivorous habits by Ephippidae species

In this section, original data on four Ephippidae species are compared with information from the literature, to clarify the importance of herbivory for the species studied. Point surveys were held in Japanese and Brazilian waters, in order to compare diet and feeding habits of the Ephippidae occurring in the Pacific and Atlantic. Sampling was due to mid-summer 2005 to early winter 2006, in Japan, and from early summer 2008 to mid-winter 2010 in Brazil.

3.1. Sampling sites

Field sampling activities were held in the reefs off the Okinawa Archipelago, Japan (JPN); and in four sites along the Western Atlantic, all in the Brazilian coast (BR) (Fig. 1). Methods for capturing fish samples varied according to the surveyed location, using nets, spearfishing, line and hook, and direct acquisition from employed fishermen or from local markets. In Brazil, most sampling sites consisted in estuarine environments (Curuçá, Bragança and some samples from Caravelas). The Table 2 summarizes sampling activities in each of the surveyed locations, detailing respective methodologies as employed.

Sampling site	Taxon	SL range (cm)	N	Methods
Okinawa (JPN)	*P. boersii*	8.54 ± 1.07	17	Employed fishermen, using several net types and hook-and-line; local fish market for all samples from Japanese waters
	P. pinnatus	18.33 ± 2.92	3	
	P. teira	17.20 ± 2.42	3	
Curuçá (BR)	*C. faber*	18.61 ± 1.14	33	Cast nets; gill nets, hook-and-line
Bragança (BR)	*C. faber*	15.42 ± 0.91	56	Local fish market
Natal (BR)	*C. faber*	9.03 ± 1.73	4	Hook-and-line
Caravelas (BR)	*C. faber*	11.86 ± 1.02	42	Gill nets; spearfishing; hook-and-line
TOTAL	-	-	158	-

Table 2. The four analyzed Ephippidae species, with regards to the methods as employed due to each sampling site. Standard size (SL) is provided as average values ± standard deviation. Wherever spearfishing was employed, samples were primarily targeted in the head, in order to cause the less damage to the stomach contents as possible.

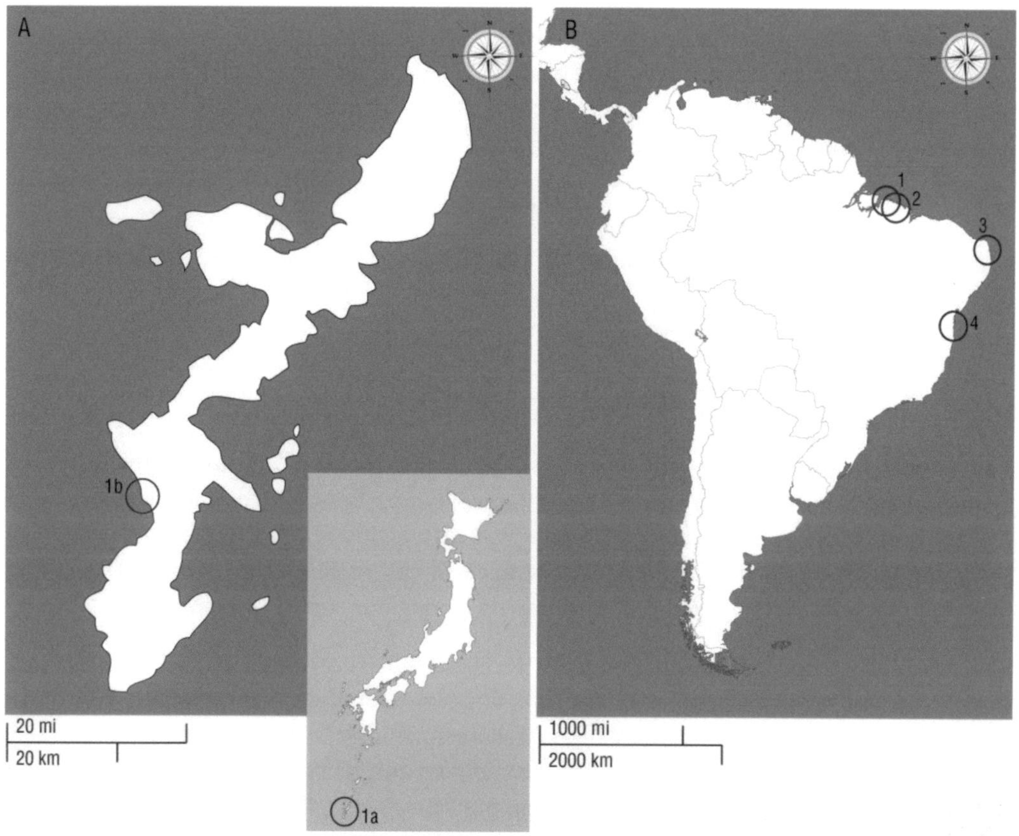

Figure 1. Sampling sites for the present study, where in (A) is the city of Chatan, Okinawa, Japan (1a-b); and (B) are sites as sampled in Brazil, in the cities of (1) Curuçá, (2) Bragança, (3) Natal and (4) Caravelas, in the states of Pará (1-2), Rio Grande do Norte (3) and Bahia (4), respectively.

3.2. Stomach contents analysis

Samples were placed in ice soon after capture, and stocked frozen until analysis, when we proceeded with dissection of stomachs, by cutting above the cardiac sphincter (esophagus) and below the pyloric sphincter (large intestine). An incision was made along the longitudinal axis, with contents removed with pincers, followed by rinsing the inner cavity with 70% ethanol. After rinsing and sorting, contents were identified to the lowest possible taxon. We proceeded with the analysis, using the protocol as adapted from Lima-Junior & Goitein (2001), which consists basically in calculating an index for Absolute Importance (AI) for each food item (*i*) present in a given sample.

The absolute importance index AI*i* was calculated for each food item by multiplying the frequency of occurrence F*i* (%) by the volumetric analysis index Vi observed in the fish diet. F*i* (%) was obtained by the formula

$$Fi\ (\%) = 100n_i/n$$

where n_i was the number of stomachs filled with food item i, and n the total number of stomachs sampled. Vi was given by the standard weight of all samples. For each food item i, points were ascribed using integers according to the degree of fullness and degree of importance of food item i, following the formula

$$Mi = \sum i/n$$

where Mi is the mean of ascribed points for i. After assessing Mi values, Vi can be calculated as

$$Vi = 25Mi$$

where 25 is a constant of multiplication.

The results were compared with those available in the literature, to any of the surveyed species, plus relevant data on herbivory activities by any Ephippidae. Information on diet and feeding habits by juvenile *P. orbicularis* was also included, due to the particularity of switching feeding behaviors from herbivory to carnivory within a day (Barros et al., 2008), yet adult fish were neither observed nor sampled during our survey in Japan.

Our results for both *Platax* and *Chaetodipterus* are summarized in Table 3. The main food item found in the stomach contents of all 23 *Platax* individuals was green algae, found in > 90% of the stomachs, with extremely high AI values, contrast to what is generally expected, as all targeted species are supposed to feed mainly on animal prey (Hayse 1990, Kuiter and Debelius 2001, Randall 2005a). Instead, animal prey consisted only in a minor food items, all with lower AI values. Similarly, *C. faber* presented green algae was the most frequently observed food item; in the three sampling sites surveyed (average AI = 2741.87 ± 83.92), yet AI levels were slightly lower than those observed for *Platax* fish; and animal food was observed more often, with substantive AI values, considering all-pooled data. However, underwater *in situ* observations using SCUBA at Caravelas suggest consumption of green algae by *C. faber* as being an incidental first step when reaching for benthic prey sheltered in the algae cover (Barros et al., in preparation). Furthermore, benthonic prey, mainly unidentified gastropod shell fragments, were observed for both genera in considerable minor frequencies rather than other food items.

The greatest evidence for feeding on benthic animal prey were shell fragments, polychaetes and bryozoa, frequently observed in the stomach contents of all four species. Sand fragments as found in stomachs of both Brazilian and Japanese samples would also indicate feeding on a benthic environment, but it is not necessarily an excluding factor, as sand grains occasionally occur on weeds from shallow or turbulent environments.

Our data contrast with the most as available in the literature regarding trophic classification of the Ephippidae. While most of the literature dealing with *Platax* species classifies all five species as carnivores (Table 1), the diet of all three *Platax* species studied in Japan is mostly algae (Table 3). Juvenile *P. orbicularis* were also observed feeding predominantly on algae turfs from hard substrates in high frequencies during the daylight, and were mostly zooplanktivores at night (Barros et al., 2008). Regardless, all available literature on *Platax* fishes refers to carnivorous habits combined with other food sources (Table 1). The Atlantic spadefish *C. faber* is referred as mainly carnivore by Hayse (1990), yet previous studies show different dietary patterns, relying on herbivory (Bittencourt, 1980; Couto & Vasconcelos Filho, 1980). Our results present a more plastic diet for this particular species, making it difficult to predict which food source is predominant for its trophic classification.

4. Functional role vs. feeding plasticity in Ephippidae

4.1. Effects of herbivore functional groups on habitat

Herbivory by fishes is a wide-ranging subject, and aspects of behavioral ecology and diet have been addressed since early records (Hiatt & Stratsburg, 1960; Jones, 1968; Borowitzka, 1981; Lewis, 1985), whereas functional ecology has been approached more recently (Bellwood et al., 2002; Bonaldo et al., 2011; Kopp et al., 2012). Efforts concerning herbivory by fishes have been made available in the literature, especially those regarding herbivory in recovery from damage (Bellwood et al., 2004; Ctanovic & Bellwood, 2009; Green & Bellwood, 2009). Intense herbivory activity reduces competition for space between corals and algae, herbivorous fish are widely recognized as a critical functional group on coral reefs. Herbivore fish families such as Acanthuridae and Scaridae are most prominent in this functional group, due to their broad distribution over tropical regions and their dense populations in such habitats (Francini-Filho et al., 2010), although many other fish groups dwelling coral reef environments may play a similar role (Ctanovic & Bellwood, 2009).

4.2. Herbivory by Ephippidae

General biology of ephippid fish is still a matter of controversy. Juveniles of the most species are cryptic, usually mimetic, dwelling coastal environments, and generally solitary, whereas adults usually form huge shoals, migrating over long distances, in up to 30m deep environments (Kuiter & Debelius, 2001; Nakabo, 2002). While there are a few reports focusing on juvenile mimic biology (Breder, 1942; Randall, 2005b; Barros et al., 2008; 2011), studies concerning adult biology, especially behavioral ecology, are rare. Why do adult fish form such shoals is therefore unknown, yet some authors suggest migratory shoaling as for reproductive reasons (Kuiter & Debelius, 2001). In coastal environments, such as coral reefs and estuaries, late juveniles and adults of the genera *Platax* and *Chaetodipterus* are usually observed solitary, in pairs or small aggregates (B. Barros, personal observation; Bellwood et al., 2006; Ctanovic & Bellwood, 2009), while mimic juveniles are usually observed solitary (Barros et al., 2008).

Species	Sampling Site (N)	Main food items observed	F*i* (%)	V*i*	AI*i*
P. boersii	OKN (17)	Algae	94.17	35.60	3352.86
		Copepoda	23.66	51.79	1225.47
		Mysida	15.92	44.86	714.28
		Teleostei	6.11	18.83	115.06
		Sand fragments	48.79	41.35	2017.51
		Und.	38.26	10.93	418.09
P. pinnatus	OKN (3)	Algae	92.25	42.32	3904.14
		Pine seed	33.33	55.66	1855.06
		Copepoda	18.92	29.84	564.67
		Gastropoda (shell fragments)	23.42	5.18	121.45
		Sand fragments	73.04	12.93	944.78
P. teira	OKN (3)	Algae	90.81	54.53	4952.14
		Sand fragments	75.00	13.63	1022.03
		Und.	11.41	22.43	255.94
C. faber	CUR (33)	Algae	56.95	43.93	2501.81
		Porifera	21.73	6.52	141.67
		Bryozoa	13.04	3.26	42.53
		Polychaeta	13.04	29.56	386.56
		Annelida	13.04	3.26	141.67
		Bivalvia	17.39	4.34	75.47
		Teleostei	13.04	7.60	99.24
		Sand fragments	56.52	34.62	1956.72
		Und.	35.28	14.55	513.32
C. faber	BRA (56)	Algae	60.23	52.50	3162.07
		Hydrozoa	6.03	10.75	64.82
		Bryozoa	18.00	30.50	549
		Oligochaeta	2.13	0.50	1.065
		Polychaeta	44.20	22.50	994.50
		Bivalvia	2.00	0.75	1.50
		Crustacea	20.05	15.00	300.75
		Teleostei	58.12	12.15	706.52
		Und.	43.94	66.14	2906.19
C. faber	NAT (4)	Algae	53.92	31.55	1701.18
		Porifera	38.60	17.91	691.32
		Polychaeta	45.73	39.55	1808.62
		Und.	67.03	35.12	2354.09
C. faber	CAR (42)	Algae	93.11	38.69	3602.42
		Polychaeta	77.68	18.04	1401.41
		Crustacea	47.85	22.34	1069.16
		Copepoda	32.23	29.81	960.79
		Isopoda	32.30	20.68	667.94
		Teleostei	7.38	17.17	126.72
		Sand fragments	73.15	24.56	1796.67
		Und.	84.25	35.87	3022.04

Table 3. Stomach contents as observed in the four locations analysed, where F*i* means "Frequency of occurrence"; V*i* means "Volumetric Index"; AI*i* means "Absolute Importance Index"; Und. means "undeterminated

Gerking (1994) has stated that no adult herbivorous fish are obligate plant eaters, selectively excluding all animal food from the diet, as larval herbivorous fish are often recorded to feed on algae plankton for an initial short period, then switching into zooplanktivory. After having developed all morphological and physiological characters, fish do shift back into herbivorous habits. However, for those groups usually classified as herbivorous, animal food items in the gut contents are considered rare, and often referred as incidentally ingested while fish are grazing. Despite our present results suggest herbivory as a major foraging tactic for all analyzed species, considering all-pooled data, animal protein input is still as important as algae ingestion, especially for *C. faber* sampled in the Brazilian coast. Although potentially eligible as a "sleeping functional group", the combination of herbivory and recovery of phase-shifted corals may be independent phenomena, as observed by Bellwood et al. (2006), when three adult *P. pinnatus* were observed foraging on a substantial layer of *Sargassum* algae from whitened coral reef in a considerably short time, when major herbivory activity by other fish groups was expected.

Although no field observations of feeding were made in Japan, our results for *Platax* from the Okinawan archipelago might be subjected to a similar phenomenon as that observed for *P. pinnatus* in the great reef barrier, considering the actual status of the reefs in the Okinawa archipelago, which at least two mass bleaching events were registered for the last ten years due to the global seawater warming, and several coral reefs colonies have been reported to experience phase shift (Loya et al., 2001, Nadaoka et al., 2001, Suefuji & van Woesik 2001, Bena et al., 2004). Algal ingestion by *C. faber* as observed in Brazil, however, might solely indicate feeding plasticity in both diet and behavior. Moreover, feeding plasticity in ephippid fishes may also be a strategy developed during early growth stages, as reported for juvenile *P. orbicularis* (Barros et al., 2008; 2011) The invasive potential of some species may indicate that the group is highly adaptable to novel food sources (Bilecenoglu & Kaya, 2006; Golani et al., 2011). In such cases, a plastic diet combined with plastic feeding behavior would favor the group while migrating to new areas.

5. Conclusions

Even for a limited number of individuals for both genera, our results suggest herbivory as the main feeding habits of ephippid fish, conflicting with the reports of a more carnivorous diet. Unless batfishes and spadefishes have been misclassified as carnivores, our data seems to be exceptional. Our results as presented here, supported by morphological data (Gregory, 1933) and behavioral data on both *Platax* (Barros et al., 2008) and *Chaetodipterus* (Barros et al., in preparation) indicate strong evidence of diverse dietary patterns, where plant material plays a major role.

Despite adult *C. faber* having a more plastic diet, herbivore habits definitely figure among the main strategies as used by that species, even if incidentally while reaching for benthic prey (Barros et al., in preparation). Adult batfishes of genus *Platax*, however, presented a consistent pattern where herbivory figures undoubtedly among the main energy input, with AI values three times as higher as all other food categories observed. Even so, *Platax* species are often observed solitary, in pairs, or small groups in shallow coral reef environments (Ctanovic & Bellwood, 2009; B. Barros, per. obs.). A bigger population would be expected for larger impacted areas, with phase-shifted corals. Conversely, *Chaetodipterus* are more common in estuarine environments (Heemstra, 2001), being particularly rare in the adjacent coral reefs of the surveyed area in the Brazilian coast (B. Barros, per. obs.).

To corroborate the predictions of Bellwood et al. (2006), we strongly recommend further *in situ* investigations focusing on foraging activities of both *Platax* and *Chaetodipterus* in reefs in the Pacific and Atlantic, as any other Family potentially eligible as functional groups occasionally dwelling both areas.

Author details

Breno Barros
Universidade Federal do Pará, Instituto de Estudos Costeiros, Brazil
Hiroshima University, Graduate School of Biosphere Science, Department of Bioresource Science, Laboratory of Aquatic Resources, Japan

Yoichi Sakai, Hiroaki Hashimoto and Kenji Gushima
Hiroshima University, Graduate School of Biosphere Science, Department of Bioresource Science, Laboratory of Aquatic Resources, Japan

Yrlan Oliveira, Fernando Araújo Abrunhosa and Marcelo Vallinoto
Universidade Federal do Pará, Instituto de Estudos Costeiros, Brazil

Acknowledgement

Thanks are due to Y. Masui (Blue-7-Sea) and the Chatan fishermen cooperative, in Okinawa, Japan; the fishermen community of the Cassurubá extractivist reserve, in Bahia, Brazil. We are also thankful to all members of the Laboratory of Aquatic Resources), especially S. Kamura (Hiroshima University) U. Scofield and L. Rabelo (CEPENE); J. Neto (ICMBio); R. L. Moura, R. Francini-Filho and E. Marocci (Conservation International) in Caravelas, Brazil; J. Meirelles and C. Cardoso (I. Peabiru) in Curuçá, Brazil, due to all logistic support. We are deeply grateful to H. Higuchi (MPEG) for reviewing and proofreading the manuscript. Thanks are also due to the Brazilian Council for Research and Development (CNPq, Grant #141225/2008-4), the Brazilian Federal Agency for Support and Evaluation of Graduate Education (CAPES, Process #6718-10-8), and the Ministry of Education, Culture, Sports Science and Technology of Japan (MEXT) for financial support.

6. References

Allen, G. R. (1981). Ephippidae. *In:* Fischer, W.; Bianchi, G. & Scott W.B. (eds.) *FAO species identification sheets for fishery purposes. Eastern Central Atlantic. (Fishing Areas 34, 47).* Vol. 2. FAO, Rome

Barros, B.; Sakai, Y. ; Hashimoto, H. & Gushima, K. (2008). The feeding behaviors of leaf-like juvenile *Platax orbicularis* (Ephippidae) observed at Kuchierabu-Jima Island, Southern Japan, *Journal of Ethology*, 26, 287-293

Barros, B.; Sakai, Y. ; Hashimoto, H. & Gushima, K. (2011). Effects of prey density on nocturnal zooplankton predation throughout the ontogeny of juvenile *Platax orbicularis* (Teleostei: Ephippidae). *Environmental Biology of Fishes*. 91 (2): 177-183

Bellwood, D. R.; Hughes, T. P.; Folke, C. & Nystrom, M. (2004). Confronting the coral reef crisis. *Nature* 429, 827–833

Bellwood, D. R.; Hughes, T. P. & Hoey, A. S. (2006). Sleeping Functional Group Drives Coral-Reef Recovery. *Current Biology*, 16 (24): 2434-2439

Bellwood, D. R.; Wainwright, P. C.; Fulton, C. J. & Hoey, A. (2002). Assembly rules and functional groups at global biogeographical scales. *Functional Ecology* 16: 557-562

Bena, C. & van Woesik, R. (2004). The impact of two bleaching events on the survival of small coral colonies (Okinawa, Japan). *Bulletin of Marine Science*, 75:115-125

Bilecenoglu, M. & Kaya, M. (2006). A new alien fish in the Mediterranean sea – *Platax teira* (Forsskål, 1775) (Osteichthyes: Ephippidae). *Aquatic Invasions*,1, (2): 80-83

Bittencourt, M.L. (1990). Preliminary investigations about trophic relations of Atlantic-spadefish *Chaetodipterus faber* (Broussonet, 1782), (Pisces, Ephippidae) in the Guaraquecaba Bay, estuarine complex of Paranagua (Parana State, Brazil). *Brazilian Archives of Biology and Technology*, 33 (1): 195-203

Bonaldo, R. M.; Krajewski, J. P. & Bellwood, D. R. (2011). Relative impact of parrotfish grazing scars on massive Porites on Lizard Island, Great Barrier Reef. *Marine Ecology Progress Series* 423: 223-233

Borowitzka, M. A. (1981). Algae and grazing in coral reef ecosystems. *Endeavour* 5(3): 99-106

Breder, C. M. (1946). An analysis of the deceptive resemblances of fishes to plant parts, with critical remarks on protective coloration, mimicry and adaptation. *Bulletin of the Bingham Oceanographic Collection* 10, 1-49

Burgess, W. & Fischer, W. (eds) (1978). *FAO species identification sheets for fishery purposes. Western Central Atlantic (Fishing Area 31)*, Vol. 2, FAO, Rome

Choat, J. H.; Robbins, W. D. & Clements, K. D. (2004). The trophic status of herbivorous fishes on coral reefs II. Food processing modes and trophodynamics. *Marine Biology*, 145 (3): 445-454

Couto, L. M. M. R. & Vasconcelos Filho, A. L. (1980). Estudo ecológico da região de Itamaracá, Pernambuco - Brasil. VIII sobre a biologia de *Chaetodipterus faber* (Broussonet, 1782) Pisces – Eppiphidae, no Canal de Santa Cruz. *Trabalhos de Oceanografia da Universidade Federal de Pernambuco*, 15: 311-322

de la Cruz-Agüero, J.; Arellano Martínez, M.; Cota Gómez, V. M. & de la Cruz-Agüero, G. (1997). *Catalogo de los peces marinos de Baja California Sur*. IPN-CICIMAR, La Paz, Mexico.

Ctanovic, C. & Bellwood, D. R. (2009). Local variation in herbivore feeding activity on an inshore reef of the Great Barrier Reef. *Coral Reefs*, 28: 127-133

Desoutter, M. (1990). Ephippidae. *In:* Quero, J. C.; Hureau, J. C.; Karrer, C.; Post, A. & Saldanha, L. (eds.) *Check-list of the fishes of the eastern tropical Atlantic (CLOFETA)* Vol. 2. JNICT, Lisbon; SEI, Paris; and UNESCO, Paris

Diana, J. S. (1995). *Biology and ecology of fishes*. Biological Sciences, London

Fischer, W.; Sousa, I.; Silva, C.; de Freitas, A.; Poutiers, J.M.; Schneider, W.; Borges, T.C.; Feral, J.P. & Massinga, A. (1990). *Fichas FAO de identificaçao de espécies para actividades de pesca. Guia de campo das espécies comerciais marinhas e de águas salobras de Moçambique.* FAO, Roma

Francini-Filho, R. B.; Ferreira, C. M.; Coni, E. O. C.; Moura, R. L. & Kaufman, L. (2010). Foraging activity of roving herbivorous reef fish (Acanthuridae and Scaridae) in eastern Brazil: influence of resource availability and interference competition. *Journal of the Marine Biological Association of the United Kingdom*, 90(3), 481–492

Gerking, S. D. (1994). *Feeding ecology of fish*. Academic, London

Golani, D.; Sonin, O. & Edelist, D. (2011). Second records of the Lessepsian fish migrants *Priacanthus sagittarius* and *Platax teira* and distribution extension of *Tylerius spinosissimus* in the Mediterranean. *Aquatic Invasions*, 6 (S1), 7-11

Green, A. L. & Bellwood, D. R. (2009). *Monitoring functional groups of herbivorous fishes as indicators of coral reef resilience – A practical guide for coral reef managers in the Asia Pacific Region*. IUCN, Gland, Switzerland. 70p.

Gregory, W. K. (1933). Fish skulls – A study of the evolution of natural mechanisms. *Transactions of the American Philosophical Society*, 23 (2): 75-481

Hayse, J. W. (1990). Feeding habits, age, growth and reproduction of atlantic spadefish *Chaetodipterus faber* (Pisces: Ephippidae) in South Carolina. *Fisheries Bulletin*, 88: 67-83

Heemstra, P. C. (2001). Ephippidae - Spadefishes (Batfishes). *In*: Carpenter, K. E, & Niem, V. (eds) *FAO species identification guide for fishery purposes. The living marine resources of the western central Pacific*, vol 6. FAO, Rome

Hiatt, R. W. & Stratsburg, D. W. (1960). Ecological relationships of the fish fauna on coral reefs of the Marshall Islands. *Ecological Monographs*, 30: 65-127

Holcroft, N. I. & Wiley, E. O. (2008). Acanthuroid relationships revisited: a new nuclear gene-based analysis that incorporates tetraodontiform representatives. *Ichthyological Research*, 55 (3): 274-283

Humann, P. & DeLoach, N. (2006). *Reef fish identification – Florida, Caribbean, Bahamas, 3rd edition*. New World Publications

Jones, R. S. (1968). Ecological relationships in Hawaiian and Johnston Island Acanthuridae (surgeonfishes). *Micronesia*, 4: 309-361

Kishimoto, H.; Hayashi, M.; Kohno, H. & Moriyama, O. (1988). Revision of japanese batfishes, genus *Platax*. *Scientific Reports of Yokosuka City Museum* 36: 19–40

Kopp, D.; Bouchon-Navaro, Y.; Louis, M.; Legendre, P. & Bouchon, C. (2012). Spatial and Temporal Variation in a Caribbean Herbivorous Fish Assemblage. *Journal of Coastal Research*, 28 (1A): 63-72

Kuiter, R. & Debelius, H. (2001). *Surgeonfishes, Rabbitfishes and their relatives: A comprehensive guide to Acanthuroidei*, TMC publishing, Chorleywood, UK

Kuronuma, K. & Abe, Y. (1986). *Fishes of the Arabian Gulf*. Kuwait Institute for Scientific Research, State of Kuwait

Lewis, S. M. (1985). Herbivory on coral reefs: algal susceptibility to herbivorous fishes. *Oecologia* 65: 370-375

Lieske, E. & Myers, R. (1994). *Collins Pocket Guide. Coral reef fishes. Indo-Pacific & Caribbean including the Red Sea*. Haper Collins Publishers.

Lima-Junior, S. E. & Goiten, R. (2001). A new method for the analysis of fish stomach contents. *Acta Scientiarum* 23:421–424

Loya, Y.; Sakai, K.; Yamazato, K.; Nakano, H.; Sambali, H. & van Woesik, R. (2001). Coral bleaching: The winners and losers. *Ecology Letters*, 4: 122-131

Masuda, H.; Amaoka, K., Araga, C.; Uyeno T.; & Yoshino, T. (1984). *The fishes of the Japanese Archipelago*. Vol. 1. Tokai University Press, Tokyo, Japan.

Maugé, L.A. (1984). Ephippidae. In Fischer, W. & Bianchi, G. (eds.) *FAO species identification sheets for fishery purposes. Western Indian Ocean (Fishing area 51)*. Vol. 2, FAO, Rome.

Myers, R. F. (1991). *Micronesian reef fishes*. Second Ed. Coral Graphics, Barrigada, Guam

Nadaoka, K.; Nihei, Y.; Wakaki, K.; Kumano, R.; Kakuma, S.; Moromizato, S.; Omija, T.; Iwao, K.; Shimoike, K.; Taniguchi, H.; Nakano, Y. & Ikema, T. (2001). Regional variation of water temperature around Okinawa coasts and its relationship to offshore thermal environments and coral bleaching. *Coral Reefs*, 20: 373-383

Nakabo, T. (ed.) (2002). *Fishes of Japan with pictorial keys to the species - English edition*. Tokai University Press, Tokyo

Nanake, Y.; Suda, Y. & Sano, M. (2011). Food habits of fishes on an exposed sandy beach at Fukiagehama, South-West Kyushu Island, Japan. *Helgoland Marine Research* 65: 123–131

Nanjo, K.; Kohno, H. & Sano, M. (2008). Food habits of fishes in the mangrove estuary of Urauchi River, Iriomote Island, southern Japan. *Fisheries Science* 74: 1024–1033

Nelson J (ed.) (2006). *Fishes of the world*, Wiley press, NY

Randall, J. E. (2005a). *Reef and shore fishes of the south Pacific: New Caledonia to Tahiti and the Pitcairn Islands*. University of Hawaii Press, Hawai

Randall, J. E. (2005b). A review on mimicry in marine fishes. *Zoological Studies*, 44 (3): 299-328

Robins, C. R.; Bailey, R. M.; Bond, C. E.; Brooker, J. R.; Lachner, E. A.; Lea, R. N. & Scott, W. B. (1991). *World fishes important to North Americans. Exclusive of species from the continental waters of the United States and Canada*. American Fisheries Society Special Publications

Russo, T.; Pulcini, D.; O'Leary, Á.; Cataudella, S. & Mariani, S. (2008). Relationship between body shape and trophic niche segregation in two closely related sympatric fishes. *Journal of Fish Biology*, 73:809–828

Schneider, M. (1995). Ephippidae. Pegualas, curacas. *In*: Fischer, W.; Krupp, F.; Schneider, W.; Sommer, C.; Carpenter, K.E. & Niem, V. (eds.) *Guia FAO para Identification de Especies para lo Fines de la Pesca. Pacifico Centro-Oriental*. FAO, Rome

Suefuji, M. & van Woesik, R. (2001). Coral recovery from the 1998 bleaching event is facilitated by Stegastes (Pisces: Pomacentridae) territories, Okinawa, Japan. *Coral Reefs*, 20: 385-386

Chapter 4

The Study of Herbivory and Plant Resistance in Natural and Agricultural Ecosystems

Michael J. Stout

Additional information is available at the end of the chapter

http://dx.doi.org/10.5772/51990

1. Introduction

The successful perpetuation of an arthropod herbivore for part or all of its life cycle on a plant- the use of a plant as a host - is typically the result of a complex and multifaceted process. At each step in the process, the herbivore interacts not only with the potential host plant but also directly or indirectly with other organisms at the same trophic level, such as competing herbivores, and with organisms at different trophic levels, such as predators and parasitoids. A great diversity of plant traits may affect these interactions and, moreover, different plant traits may be relevant at different steps in the process; visual and odor cues emitted by the plant, for example, may be used by herbivores (and natural enemies) for long- or mid-range host location, whereas non-volatile secondary chemicals may be involved in the process only after the herbivore begins to feed on the plant. Any plant trait that varies among individual plants and that affects an aspect of the herbivore's interaction with the plant or with other organisms associated with the plant is potentially a basis for differences among plants in the level of damage caused by the herbivore (i.e., plant resistance). Thus, the study of plant resistance involves the study of a large web of interactions mediated by a potentially large and diverse set of plant traits, and plant resistance can be studied from various perspectives.

Over 25 years ago, Kogan [1] noted the existence of two parallel bodies of research and theory related to the study of plant resistance. The first, which he referred to as the "Insect-Plant Interactions" (IPI) literature, was concerned with describing and explaining the ecological and evolutionary relationships among the two most diverse groups of terrestrial organisms, with a particular emphasis on explaining patterns of variation in the expression of resistance-related traits among plants. The second, which he termed the "Host-Plant Resistance" (HPR) literature, was the province of practically oriented scientists concerned with the development and deployment of crop varieties resistant to herbivores. As these two bodies of research and theory deal with similar biological phenomena, researchers in the two fields have much to learn from one another. Historically, however, communication between these two groups has been only partial. A great many advances have been made in both HPR and IPI in the years since Kogan's review, but barriers to the exchange of ideas and data among these two groups of scientists still exist, and reconciliation of the two literatures has not yet been completed.

Reconciliation of the HPR and IPI literatures is a matter of considerable practical as well as academic interest. Insect pests significantly reduce the yield and quality of all major plant commodities [2], and the use of insecticides to control insect pests is attended by numerous problems, including high costs associated with both product and product application, elimination of populations of natural enemies, development of insecticide resistance and resurgence by target pests, insecticide-induced emergence of destructive secondary pests, and negative impacts of insecticides on human health and the environment [3]. Increased understanding of the ecology of plant-insect interactions and the proper application of this understanding to crop-pest interactions has led in the past, and will lead in the future, to more effective, less damaging means of managing pests, including the development of resistant crop varieties. The purpose of this chapter is to draw renewed attention to the problematic relationship between the IPI and HPR literatures and to the barriers to the exchange of ideas and data among the two literatures. To that end, I will proceed by first considering the conceptual foundations of IPI and HPR research and then by considering the categorical frameworks under which research in IPI and HPR is conducted. I will conclude with suggestions for applying insights and advances from the IPI literature over the past few decades to HPR research.

2. The conceptual foundations of IPI and HPR research

The seminal work in the establishment of HPR as a distinct discipline was Reginald Painter's *Insect Resistance in Crop Plants*, first published in 1951 [4]. *Insect resistance* is striking for its sophisticated understanding of the complexities of crop plant-pest interactions and for its prescience (with respect to the latter point, the importance to plant resistance of plant phenotypic plasticity and plant tolerance were both points made by Painter but not picked up in the IPI literature until later). Painter's book is also striking because it is bereft of connections to broader ecological or evolutionary theory beyond general applications of the principle of natural selection. Following Painter's lead, most HPR research has retained a heavily empirical and practical orientation, typified by the following statement of Painter's: "The agronomist does not demand a full knowledge of the causes of high yield before breeding for this character in field crops. It is no more necessary to know the exact cause in breeding for insect resistance." [4, pg. 75]

The IPI literature, in contrast, is characterized by a rich tradition of generating and testing hypotheses designed to explain patterns in plant-insect interactions. Although many papers played important roles in establishing the discipline of IPI, two of undoubted importance were those by Fraenkel in 1959 [5] and Ehrlich and Raven in 1964 [6-9]. The paper by

Fraenkel established the focus in the IPI literature on secondary plant metabolites as the primary mediators of plant-insect interactions and also contained inchoate ideas of reciprocal evolutionary relationships among plants and plant-feeding insects [7, 10]. Ehrlich and Raven further developed the concept of plant-insect coevolution in which herbivores and plants were viewed as important drivers of one another's evolution. These authors argued that the fitness-reducing effects of herbivores on plants has selected for the evolution of novel defensive traits in plants, and that the possession of effective defenses by plants has selected for the evolution by insects of adaptations allowing them to overcome these novel plant traits. The evolution of countermeasures to plant defenses by herbivores has acted, in turn, as a selective pressure for the development of further plant defenses, and so on in an escalating reciprocal fashion. According to Ehrlich and Raven, this coevolutionary arms race involving "novel defensive breakthroughs" in plants and "offensive innovations" by herbivores [11] has shaped patterns of variation in plant defense and has served as an important impetus for specialization and diversification in both herbivorous insects and plants.

The ideas of Fraenkel and Ehrlich and Raven have proven to be very fertile and have spawned a number of more specific hypotheses designed to explain patterns of variation in expression of plant defenses at various taxonomic, spatial, and temporal scales [7]. The most influential of these hypotheses have been the optimal defense hypothesis, the growth rate hypothesis, the carbon:nutrient balance hypothesis, and the growth-differentiation balance hypothesis [8, 12]. According to the optimal defense hypothesis, plant defenses at different spatial and taxonomic scales are allocated in a manner that optimizes plant fitness by minimizing the costs and maximizing the benefits of defense expression. The carbon:nutrient balance hypothesis views phenotypic variation in allocation to plant defense as a result of the supply of carbon and nutrients (primarily N) in the environment. The growth rate hypothesis focuses on inherent plant growth rate, itself determined in evolutionary time by resource availability, as the most important determinant of investment in plant defense. Finally, the growth-differentiation balance hypothesis views allocation to plant defense in light of a tradeoff between plant growth and differentiation. Stamp [8] provides a thorough overview of these hypotheses.

Various revisions of these ideas have of course been made in the past five decades. The importance of plant primary metabolites and morphological traits for plant defense has been recognized (10). Also, it has become apparent that the defensive phenotypes of most plants have been shaped by the need to defend against multiple types of attackers simultaneously and thus "diffuse" coevolution is probable more common than the "escape-and-radiate" or "pairwise" coevolution envisioned by Ehrlich and Raven and others [10,11,13]. It has also become apparent that the influence of plant defenses on insect evolution has probably been stronger than the influence of insects on plant evolution [11]. Furthermore, experimental support for all of the specific hypotheses developed to explain patterns of plant defense allocation has been equivocal; although none of these hypotheses has been fully rejected, none of them provides the level of generality desired and enthusiasm for testing these hypotheses has flagged somewhat in recent years [7,8]. These various revisions and developments notwithstanding, the overall paradigm of a coevolutionary arms race between plants and herbivores mediated largely by plant secondary metabolites remains strongly entrenched as the guiding paradigm for IPI research.

Given the slim conceptual underpinning of HPR research, the real issue in reconciling the conceptual foundations of HPR and IPI research is the extent to which IPI theory is applicable to the study of crop plant-insect pest interactions. Kogan [1] appeared to believe the application of IPI theory to crop plant-pest interactions to be a relatively straightforward matter, and he presents an extended discussion of the application of optimal defense theory to crop-pest interactions. Other reviews, including reviews more recent than Kogan's, often discuss crop-pest interactions in the context of IPI theory [e.g., 14-16], even if they do not attempt detailed applications of plant defense hypotheses to crop-pest interactions.

However, there are at least two major problems with the application of IPI theory to crop plant-pest systems. The first major problem with the application of IPI theory to crop-pest systems arises from the fact that crop plants are often grown in environments very different from those present before or during the process of domestication, when coevolutionary relationships presumably developed. Crop plants are often grown in areas where they are not native, distant from their centers of origin and domestication, and thus are exposed to herbivores and other organisms with which they have no history, or only a relatively short history, of interacting. Crop plants are, in other words, exotic species in most areas where they are cultivated. Furthermore, the conditions that characterize many modern agricultural regions—large monocultures, with abundant water and high levels of fertilizer and other chemical inputs—differ from the conditions present during the process of coevolution of the crop's progenitor with herbivores. Under these circumstances, it is unclear how applicable all but the loosest notions of diffuse coevolution are to crop-pest systems, and how adapted we should expect crop plants and their insect pests to be to one another. A similar point, but applied to biological control, has been made by Hawkins et al. (17). These authors argued that biological control of pests by predators and parasitoids in crops may not reflect predator-prey interactions in natural systems, because food webs present in agricultural systems are often greatly simplified and composed largely or entirely of exotic species, and because the environments in which biological control takes place are greatly simplified in structure and ecological connectedness relative to natural systems.

The second problem with the straightforward application of IPI theory to agricultural systems arises from the fact that crop plants are domesticated, meaning their genotypes and phenotypes have been shaped not only by natural selection but also by human-guided artificial selection. Artificial selection for desired agronomic traits has quite likely altered or disrupted suites of plant resistance-related traits developed over long periods of coevolution with herbivores. This is, of course, obvious in those crop varieties that have been intentionally bred for resistance to herbivores. In these varieties, selection during breeding has resulted in the accentuation of specific resistance-related traits that reduce the impact of herbivory on crop yield, whether or not those traits are fully understood. Importantly, these resistance-related traits may or may not be the same traits favored in the absence of human action. Prominent examples of intentionally selected resistance include wheat varieties resistant to Hessian fly and maize varieties resistant to various Lepidopteran borers and defoliators; Smith and Clement (16) provide additional examples.

In addition to those cases in which crop plants have been intentionally bred for resistance, there is now ample evidence for collateral effects of selection for desired agronomic traits on crop plant resistance to insects. In some cases, the collateral effects of selection for agronomic traits on plant resistance are easily understood and somewhat predictable because they involve plant traits related to human nutrition or palatability (18). Such appears to be the case in many Solanaceous crops, in which the potential human toxicity of glycoalkaloids has led to the intentional selection of varieties with low levels of these secondary chemicals and reduced levels of resistance to some herbivores and pathogens (19). Similarly, domestication of celery has probably involved selection for reduced levels of furanocoumarins, which can have toxic and irritant effects on humans (20) but which may be involved in the resistance of celery to pests.

Probably more common are those cases in which pleiotropy and epistasis result in unintended collateral effects on plant resistance during breeding (21). Notably, in many crops, selection for increased allocation to agronomic yield and quality appears to have resulted in reduced allocation to defense. There are now a number of studies showing greater susceptibility to pests in domesticated varieties, although the precise phenotypic manifestations of this tradeoff are varied and not as yet predictable. In one of the best-studied examples, a negative relationship was found in maize between degree of domestication and defense against insects; plant growth and yields were highest but resistance to an assemblage of pests lowest in a modern maize cultivar and a land race, whereas growth and yield were lowest but pest resistance highest in annual and perennial wild *Zea* species (18). Resistance to stem borers in wild and perennial relatives of cultivated maize was attributable to greater numbers of tillers in wild varieties, which allowed the plant to compartmentalize injury by borers and thereby minimize yield losses. Wild tomato was more tolerant of defoliation than a domesticated tomato variety, possibly because of higher allocation to leaves and fruits and lower allocation to storage organs in the domesticated tomato (22). In cranberry, resistance to gypsy moth was lower on more derived, higher-yielding varieties than on wild selections (23). The reduced resistance in more derived varieties was correlated to some extent with reduced induction of sesquiterpenes and reduced levels of jasmonic acid. In sunflower, Mayrose et al. (24) found negative correlations between growth under benign environmental conditions and resistance to *Trichoplusia ni* as evidenced by greater preference for high-yielding domesticated sunflowers then for wild sunflowers. Domesticated sunflowers were also more susceptible to fungal infection and drought stress. Also in sunflower, Michaud and Grant (25) found domesticated sunflowers to be more palatable to, and more susceptible to ovipositon by, the cerambycid pest *Dectes texanus* than was a wild sunflower. The greater susceptibility of domesticated sunflower to *D. texanus* was partly attributed reduced resin flow in the domesticated variety, a trait that has been selectively diminished during breeding to facilitate harvesting.

The apparent tendency of domestication to negatively impact plant resistance to herbivores has probably been exacerbated by ignorance of mechanisms of resistance and by neglect in monitoring resistance during the breeding of most crops. Rodriguez-Saona et al. (23), for example, have pointed out that the importance of induced plant volatiles in facilitating the action of natural enemies has only recently been discovered, and the plant traits important to this mechanism of plant defense have probably been altered during selection. This idea is supported by the research of Rasmann et al . [26] who showed that the roots of maize lines developed in North America are incapable of emitting caryophellene following rootworm feeding, a lack that compromises the ability of entomopathogenic nematodes from finding and infecting rootworms. More generally, the importance of plant – natural enemy cooperation is only now being recognized, and it is very likely that these cooperative relationships have been disrupted in domesticated varieties [27]. In sunflower, abundance of sunflower moths (*Homoeosoma electellum*) was much higher and parasitism much lower on a domesticated variety than on wild sunflower. The reason for this disruption in parasitism on domesticated sunflower was related to differences in flower morphology and phenology in domesticated and wild sunflower that affected parasitoid searching behavior and access of parasitoids to larvae (28).

Thus, the processes of domestication and plant breeding have sometimes altered crop plant genotypes and phenotypes in ways that compromise or disrupt natural, co-evolved plant defense mechanisms. Moreover, the conditions under which crops are grown may preclude or interfere with natural defense mechanisms, and the status of many crop plants as exotics means interactions between crops and pests are, from an ecological and evolutionary perspective, novel. As a general implication, this brings into question the straightforward application of IPI theory to HPR research; more specific implications are discussed below.

3. Classification schemes for the study of resistance in the IPI and HPR literatures

Because there exists such a variety of ways by which plants may reduce the impacts of herbivory, HPR and IPI researchers have often found it necessary to place resistance types into categories. As might be expected of two disciplines that differ so markedly in their conceptual foundation, the categorical frameworks developed within the HPR and IPI literatures to classify types of resistance differ. Based on observations of resistance in the field, Painter introduced a three-fold scheme for "dividing" the "phenomena of resistance" (4,16). In Painter's original scheme, the term "antibiosis" was used to describe adverse effects of resistant plants on herbivore physiology and life histories such as reduced growth, survival, and fecundity. The second category, "non-preference", comprised those plant traits that affect herbivore behavior in ways that reduce the colonization or acceptance of a plant as a host. Finally, tolerance was defined as the ability of a plant to withstand herbivore injury such that agronomic yields or quality are reduced to a lesser extent than in a less tolerant plant subjected to equivalent injury. Since Painter, the use of "mechanism" to describe these terms has largely been abandoned in favor of "modality" or "category", probably in response to increased understanding of the plant traits that underlie the actual mechanisms of plant resistance. Also, in 1978, Kogan and Ortman [29] proposed replacing "non-preference" with the term "antixenosis" to emphasize the similitude of this category with the category of antibiosis. Aside from these minor modifications, however, Painter's trichotomy has been remarkably influential and is still used widely today. For example, fully half of the articles published in the "Plant Resistance" section of the *Journal of Economic Entomology* in 2011 used the terms to describe the lines or varieties under study.

The IPI literature, in contrast, has not seen the establishment of a more-or-less formal categorization scheme comparable to Painter's trichotomy. However, over the past two decades, in response to advances in the understanding of mechanisms by which plants reduce the impact of herbivores, a bifurcated scheme has emerged (Figure 1). In this scheme, the term "resistance" is used broadly to comprehend those plant traits that reduce the extent of injury done to a plant by an herbivore, where injury is understood as effects on plant physiological processes resulting from the use by an herbivore of a plant as a host (e.g., removal of photosynthate, reduction in nutrient uptake due to root feeding). The term "tolerance" encompasses those plant traits or physiological processes that lessen the amount of damage resulting per unit injury, where "damage" is to be understood primarily in terms of plant fitness. In addition, the resistance category is often further divided into "constitutive" or "inducible" and "direct" or "indirect". Constitutive plant resistance is resistance that is expressed irrespective of the prior history of the plant, whereas inducible resistance is resistance only expressed, or expressed to a greater extent, after prior injury (i.e., expression of inducible defenses is contingent on prior attack, whereas constitutive defenses are not). Direct plant resistance refers to those plant traits that have direct (unmediated) effects on herbivore behavior or biology. Indirect plant resistance, in contrast, depends for its effect on the actions of natural enemies. The best-studied examples of indirect plant defenses are volatile organic compounds and extrafloral nectaries that facilitate the activities of natural enemies [30].

In considering the relative merits of the IPI and HPR frameworks for classifying resistance, one relevant question is whether Painter's trichotomy, which has remained virtually unchanged for the past 60 years, is capable of accommodating recent advances in the understanding of the mechanisms of plant defense. Interestingly, in Painter's original discussion of mechanisms of resistance, he acknowledged the existence of plant traits that did not appear to fit into his trichotomous scheme (4, pgs. 68-70). One such trait discussed by Painter was the long husks of some corn varieties that served as a barrier to the rice weevil; another such trait was thick walls on the pods of some bean varieties that prevented the stylets of plant bugs from reaching the seeds. In the time since Painter, research has served to reinforce the remarkable diversity of plant traits capable of affecting plant-herbivore interactions and therefore capable of serving as bases of plant resistance. Many of these plant traits do not easily fit the definitions of antibiosis, antixenosis, or tolerance set forth by Painter. A few examples will suffice. Indirect plant defenses— plant traits that act by affecting the behavior of the natural enemies of herbivores— provide a set of examples of plant defenses that do not fit easily within Painter's trichotomy. Two other examples are provided by Marquis et al. (31), who showed the resistance of white oak to a leaftying caterpillar to be related to the spatial distribution of leaves in the canopy and the percentage of leaves touching on another, and Chen et al. (32), who showed that resistance of Douglas fir to the western spruce budworm was related to the phenology of bud burst. Although *ad hoc* modifications of Painter's categories can be made to accommodate these mechanisms of resistance, this cannot be done without contravening the original intent of Painter's categories.

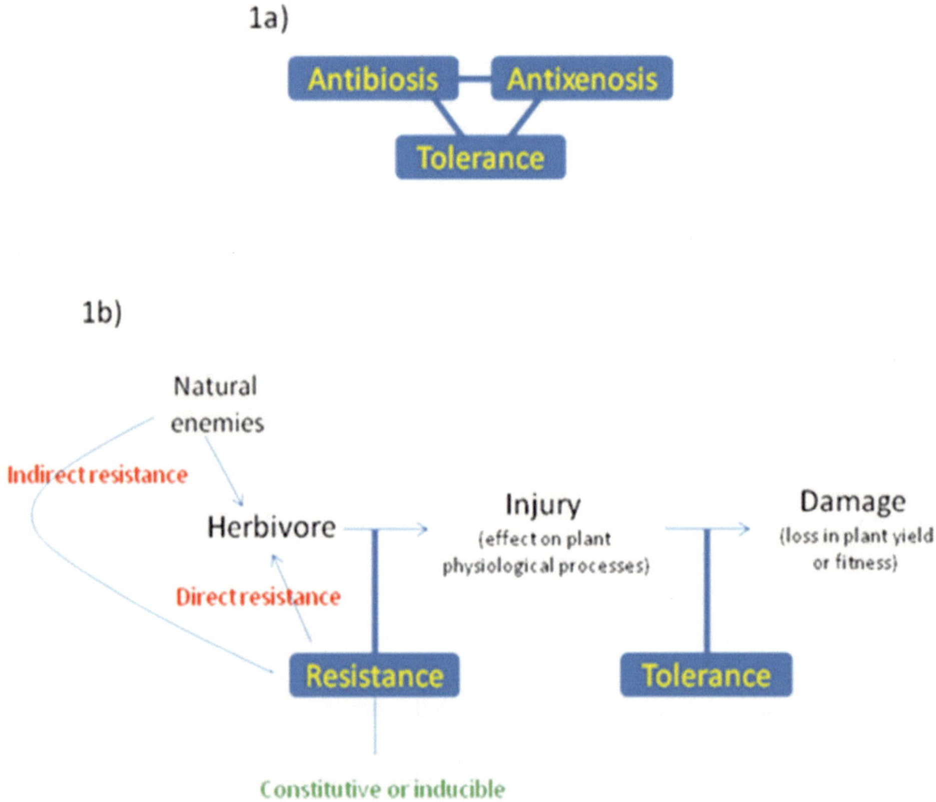

Figure 1. Comparison of the schemes used in the HPR (1a) and IPI (1b) literatures to categorize types of plant resistance to insects.

Another problematic aspect of Painter's trichotomy involves the antixenosis category and its relationship to antibiosis. In Painter's original conception, non-preference (antixenosis) was separable from antibiosis, with the two phenomena controlled by different plant genetic factors: The three [categories of resistance] are usually the result of separate genetic factors but are interrelated in their final effects…" [4, pg. 70]. There is, however, some ambiguity in Painter and in the subsequent HPR literature in the use of the terms non-preference and antixenosis. Antixenosis/non-preference has sometimes been used narrowly, to denote interference with behaviors involved in host location ("pre-alightment" behaviors). At other times, however, the term has been used very broadly to denote effects on behaviors that occur both before a potential host is located and after a potential host is located ("post-alightment" behaviors). When the term is used narrowly, non-antibiotic effects on important pest behaviors involved in the acceptance of host plants for feeding and oviposition are excluded. When the term is used broadly, various difficulties are encountered in separating antixenosis from antibiosis. This is true, for example, when resistance involves a strong element of feeding deterrence; in such cases, antixenosis can be very difficult to distinguish from antibiosis without complicated experimental procedures (33). More generally, research over the 60 years since Painter has shown that the same plant trait often, perhaps even usually, has effects on multiple aspects of a plant-insect interaction, including aspects that might be classified as both antibiotic and antixenotic. As an important example, toxic secondary chemicals are often also deterrent, and vice versa. In fact, a high degree of correspondence between deterrence and toxicity is the expectation of a facile evolutionary analysis, as insects not deterred by a toxic chemical and insects deterred by a non-toxic chemicals would seem to be at a selective disadvantage. Experimental tests of this expectation are limited, with some supporting the expectation and other not (34). As another example, the same volatile organic compound emitted by plants after herbivore feeding can attract natural enemies (indirect defense, an antibiosis-like effect), deter oviposition (antixenosis), and perhaps have direct toxic effects on insects (antibiosis) [30,35]. Antibiosis and antixenosis are hence often not separable phenomena but are effects of the same plant traits. In such cases, the results of efforts to categorize resistance often are highly dependent on experimental methods used: choice studies will reveal the resistance as antixenosis, non-choice studies, as antibiosis, even though the same plant trait is responsible for both effects.

Thus, while Painter's trichotomy has been extremely useful in advancing the science of HPR, it may not satisfactorily accomodate advances in our understanding of the mechanisms of plant resistance. Moreover, Painter's trichotomy creates a distinction (between antibiosis and antixenosis) that may not be particularly useful, since the two categories involve overlapping plant traits that have the same effect of reducing the amount of injury done by herbivores to crops. The implications of adopting the alternative categorical scheme outlined in Figure 1 are discussed below.

4. Implications

In the above, I have argued first, that the application of IPI theory to HPR may not be a straightforward matter and, second, that the categorical framework historically used by HPR researchers may not accommodate more recent advances in understanding of the mechanistic bases of plant resistance. These are not merely contrarian arguments, but have important implications for the practice of HPR. In what follows, I will seek to point out a few of the implications of these arguments and to suggest areas in need of further research.

4.1. Applying IPI theory to HPR

The issue of the extent to which coevolutionary principles and IPI theory apply to crop-pest interactions deserves more careful consideration. Perhaps the most important questions surrounding this issue relate to the effects of domestication on suites of natural (co-evolved) plant resistance mechanisms. How general is the tradeoff between crop resistance to pests and agronomic yield/quality? Such tradeoffs have now been reported in a number of crop-pest systems, but data are insufficient to conclude that tradeoffs are universal. This is particularly true if negative results (lack of a tradeoff) are less frequently reported in the literature than positive results, which seems likely. Further, when a tradeoff is present, how is the tradeoff manifested? Are certain types of resistance-related traits in plants (e.g., plant traits involved in tolerance, or traits involved in indirect defense) more likely to have been affected by domestication than others? Can patterns in the effects of domestication on resistance be discerned among different types of crops and pests? These and other questions need to be addressed to determine the extent to which domestication has altered natural suites of plant defenses, and to determine whether the effects of breeding are so far-reaching as to preclude the study of crop resistance as a natural phenomenon.

Another important question in the application of IPI theory to HPR is whether the conditions under which crop plants are typically grown make certain natural strategies of defense less effective or unavailable. For example, as Kogan [1] pointed out, the strategy of escaping injury by herbivores by being small or short lived (unapparent) may be integral to the defensive strategies of some plants (including the progenitors of some crop plants) in natural environments, but this is a strategy unavailable to crop plants grown in large monocultures . Similarly, indirect defenses, which are contingent on the activities of natural enemies, may be compromised in the simplified habitats and food webs found in agricultural fields [17].

Finally, what implications are there to the fact that crop plants are subject to attack by pests with which they share only a short history of interacting? One possible implication is that crop -pest interactions may differ qualitatively depending on whether the crop is grown near its center of domestication or elsewhere. In tropical Asia, for example, rice has a long history of domestication, probably long enough for it to co-evolve with its pests and associated organisms, and in these areas populations of many pests on locally adapted varieties are maintained below damaging levels by a combination of top-down and bottom-up factors unless disrupted by early season insecticide use or other high-input practices (36). In contrast, in temperate areas, where rice has been more recently introduced, such natural controls of pest populations appear to be more limited (Stout, personal observations). Another possible implication of the exotic nature of many crop-pest interactions relates to plant defenses triggered by the release of specific elicitors in insect oral secretions (so-called herbivore-associated molecular patterns, or HAMPs). The presence of HAMPs in the oral secretions of insects is viewed as an outcome of the "350 million-year period of coexistence, plants, insects, and other arthropods" (i.e., an outcome of coevolution) [14]. If this is the case, what level of specificity is to be expected in HAMP-mediated responses of crop plants to pests with which they share no history of coevolution?

4.2. Modifying the categorical framework of HPR research

The categorical framework under which HPR research is conducted needs to be re-examined and perhaps abandoned in favor of a scheme similar to that used in the IPI literature (Figure 1). One effect of adopting the dichotomous scheme used in the IPI literature would be to eliminate the difficulties in separating antibiosis and antixenosis. Another reason is that the "resistance" category in the IPI scheme explicitly incorporates indirect defenses and induced defenses, while Painter's trichotomy does not. In our current understanding, these are important modes or types of plant defense but are at risk of being ignored in a scheme that does not explicitly recognize them. Inclusion of indirect and induced defenses is important from a practical perspective as well. This is because these types of defenses may require the development of specific phenotyping procedures to monitor for them during the breeding process. Again, such methods are at risk of not being developed until the importance of these types of defenses is recognized.

Adoption of the scheme presented in Figure 1 may also have the salutary effect of focusing more attention on questions of relevance to pest management. As noted above, the plant traits (mechanisms) responsible for antibiosis and antixenosis often overlap, as is the case with secondary plant substances that are both deterrent and toxic. In these cases, efforts to categorize resistance as antibiosis or antixenosis may be counterproductive because they divert attention and resources from the critical question of how a particular plant trait effects a "reduction in the over-all population of the insect resisted" [4, pg. 49] by altering the biology or behavior of the pest or of other organisms associated with the plant-pest interaction. An antibiotic trait that slows the growth and development of a Lepidopteran pest may reduce pest populations to a much lesser degree than antibiotic trait that kills a large portion of early instars. Alternatively, the two antibiotic traits might bring about similar population reductions by very different mechanisms—the former trait, by direct effects on the pest; the latter trait, by synergizing the effects of natural enemies. Or, an antixenotic trait that strongly deters insect feeding, resulting in pest starvation, and an antibiotic trait that poisons a pest may reduce pest populations to similar degrees. In all these cases, the status of a trait as "antibiotic" or "antixenotic" is far less important than the mechanism by which the traits bring about reductions in pest populations the reduce injury to the crop. Recent advances in genetic manipulative techniques have made it feasible to alter plant traits with precision and to monitor the effects of such alterations on pest populations, making categorization of resistance types less important.

5. Conclusions

The promise of increasing crop plant yields and food production by developing and deploying insect-resistant crops remains partly if not largely unfulfilled [16,33]. Great technical strides have been made over the past few decades in the ability to identify and quantify secondary chemicals and other plant traits associated with plant resistance. Likewise, great strides have been made in the ability to alter expression of specific plant traits through manipulative genetic methods. These advances enable us to investigate how the presence of specific plant traits change the interactions of pests with crop plants and with associated organisms and how these changes result in reduced crop injury and damage. What is particularly needed now is an understanding of the full array of strategies by which plants lessen the impact of herbivory in natural habitats, and an understanding of how domestication and modern agronomic practices have affected this array. This task will be facilitated by the use of terminology and categories that encompass the range of strategies used by plants. Ultimately, this undertaking may allow reversal of the effects of domestication and modern cultivation practices by target breeding, genetic engineering, alteration of crop environments, and other tactics.

Author details

Michael J. Stout

Department of Entomology, Louisiana State University Agricultural Center, Baton Rouge, Louisiana, USA

6. References

[1] Kogan, M. Plant defense strategies and host-plant resistance. In: Kogan M (ed.) Ecological Theory and Integrated Pest Management Practice. New York, John Wiley and Sons; 1986. P83-134.

[2] Oerke, EC. Crop losses to pests. Journal of Agricultural Science. 2006; 144: 31-43.

[3] Ekström, G, Ekbom B. Pest control in agro-ecosystems: An ecological approach. Critical Reviews in Plant Science 2011; 30: 74-94.

[4] Painter RH. Insect Resistance in Crop Plants. Lawrence, The University Press of Kansas; 1951. All page citations are from the 1968 paperbound edition.

[5] Fraenkel GS. The raison d'être of secondary plant substances. Science 1959; 129: 1466-1470.

[6] Ehrlich PR, Raven PH. Butterflies and plants: a study in coevolution. Evolution 1964; 18: 586-608.

[7] Berenbaum MR. The chemistry of defense: Theory and practice. Proceedings of the National Academy of Science USA 1995; 92: 2-8.

[8] Stamp N. Out of the quagmire of plant defense hypotheses. The Quarterly Review of Biology, 2003; 78(1): 23-41.

[9] Berenbaum, MR, Zangerl AR. Facing the Future of Plant-Insect Interaction Research: Le Retour à la "Raison d' Être". Plant Physiology 2008; 146: 804-811.

[10] Carmona, D, Lajeunesse MJ, Johnson, M J. Plant traits that predict resistance to herbivores. Functional Ecology 2010; doi: 10.1111/j.1365-2435.2010.01794.x.

[11] Price PW, Denno RF, Eubanks MD, Finke DL, Kaplan I. Insect Ecology. Cambridge, CambridgeUniversity Press; 2011.

[12] Stamp, N. Theory of plant defensive level: example of process and pitfalls in development of ecological theory. Oikos 2003; 102: 672-678.- 677.

[13] Iwao K, Rausher MD. Evolution of plant resistance to multiple herbivores: Quantifying diffuse coevolution. The American Naturalist 1997; 149: 316-335.

[14] Mithöfer A, Boland W. Recognition of herbivory-associated molecular patterns. Plant Physiology 2008; 146: 825-831.

[15] Kim J, Quaghebeur H, Felton GW. Reiterative and interruptive signaling in induced plant resistance to chewing insects. Phytochemistry. 2011; 72: 1624-1634.

[16] Smith CM, Clement SL. Molecular bases of plant resistance to arthropods. Annual Review of Entomology 2012; 57: 309-328.

[17] Hawkins B A, Mills, N J, Jervis M A, Price P W. Is the biological control of insects a natural phenomenon? Oikos 1999; 86: 493-506.

[18] Rosenthal JP, Dirzo R. Effects of life history, domestication and agronomic selection on plant defence against insects: Evidence from maizes and wild relatives. Evolutionary Ecology 1997; 11: 337-335.

[19] Friedman M. Potato glycoalkaloids and metabolites: Roles in the plant and in the diet. Journal of Agricultural and Food Chemistry 2006; 54: 8655-8681.

[20] Diawara M, Trumble JT, Quiros CF, Hansen R. Implications of distribution of linear furanocoumarins within celery. Journal of Agriculatural and Food Cehmistry 1995; 43: 723-727.

[21] Hancock JF. Contributions of domesticated plant studies to our understanding of plant evolution. Annals of Botany 2005; 96: 953-963

[22] Welter SC, JW Stegall. Contrasting the tolerance of wild and domesticated tomatoes to herbivory: agroecological implications. Ecological Applications 1993; 3(2): 271-278.

[23] Rodriguez-Saona C, Vorsa N, Singh AP, Johnson-Cicalese J, Szendrei Z, Mescher M, Frost CJ. Tracing the history of plant traits under domestication in cranberries: potential consequences on anti-herbivore defences. Journal of Experimental Botany 2010; doi:10.1093/jxb/erq466.

[24] Mayrose M, Kane NC, Mayrose I, Dlugosch KM, Rieseberg LH. Increaed growth in sunflower correlates with reduced defences and altered gene expression in response to biotic and abiotic stress. Molecular Ecology. 2011; 20: 4683-4694.

[25] Michaud JP, Grant A K. The nature of resistance to Dectes texanus (Col., Cerambycidae) in wild sunflower, *Helianthus annuus*. Journal of Applied Entomology 2009; 133: 518-523.

[26] Rasmann S, Köllner TG, Gegenhardt J, Hiltpold I, Toepfer S, Kuhlmann U, Gershenzon J, Turlings TCJ. Recruitment of entompathogenic nematodes by insect- damaged maize roots. Nature 2005; 434: 732-736.

[27] Macfadyen S, Bohan, D A. Crop domestication and the disruption of species interactions. Basic and Applied Ecology 2010; 11: 116-125.

[28] Chen, YH, SC Welter. Crop domestication disrupts a native tritrophic interaction associated with the sunflower, *Helinathus annuus* . Ecological Entomology.2005; 30: 673-683.

[29] Kogan M, Ortman EF. Antixenosis- A new term proposed to define Painter's "nonpreference" modality of resistance. ESA Bulletin 1978; 24(2):.175-176.

[30] Degenhardt J. Indirect defense responses to herbivory in grasses. Plant Physiology 2009; 149:96-102.

[31] Marquis RJ., Lill JT, Piccinni A. Effect of plant architecture on colonization and damage by leaftying caterpillars of *Quercus alba*. – Oikos 2002; 99: 531-537.

[32] Chen C, Clancy KM Kolb TE. Variation in bud-burst phenology of Douglas-fir related to Western spruce budworm (Lepidoptera: Tortricidae) fitness. Journal of Economic Entomology 2003; 96(2): 377-387.

[33] Painter RH. Crops that resist insects provide a way to increase world food supply. Kansas State University of Agriculture and Applied Science, Agricultural Experiment Station 1968; Bulletin 520.

[34] Bernays, EA. Neural limitations in phytophagous insects: implications for diet breadth and evolution of host affiliation. Annual Review of Entomology 2001; 46: 703-727.

[35] Kessler A, Baldwin IT. Defensive function of herbivore-induced plant volatile emissions in nature. Science 2001; 291: 2141-2144.

[36] Settle, WH, Ariawan H, Astuti ET, Cahyana W, Hakim AL, Hindayana D, Lestari AS. Managing tropical rice pests through conservation of generalist natural enemies and alternative prey. Ecology 1996; 77: 1975-1988.

Chapter 5

Planned Herbivory in the Management of Wildfire Fuels

Roger S. Ingram, Morgan P. Doran and Glenn Nader

Additional information is available at the end of the chapter

http://dx.doi.org/10.5772/48673

1. Introduction

Wildfires are increasing in number, intensity, and size. Five of the most significant wildfire seasons in the United States since 1960, as measured by total area burned, have occurred since 2000 [1]. The vegetation or fuel profile, a major factor determining fire behavior, is studied in two aspects: vertical and horizontal arrangement and amount. The vertical arrangement of fuel determines the degree of its mixture with air and, thus flame height and duration of elevated heat. The continuity of horizontal fuel arrangement determines potential fire spread across the landscape. Fuel attributes, along with topography and weather conditions (wind and fuel moisture), determine the kind of wildfire that will occur. Many management and ecological conditions have allowed fuels to accumulate. The increasing number of residences occurring in forest and rangeland ecosystems provides more ignition sources and restricts the ability to manage fire. Introduction of exotic plants like cheatgrass in the Inter-Mountain region of the United States has also changed fire behavior in many sagebrush plant communities [2]. Reducing biomass and the architecture of vegetation with chemical and mechanical methods can be effective, but are costly and complicated by rough terrain. Herbivory can result in short-term seasonal impacts on vegetation amounts and structure and long-term shifts in plant community composition and structure [3]. Grazing by domesticated ruminants is perhaps the most widely applied type of herbivory and can alter vegetation to reduce wildfire risks, which is often an inadvertent result in livestock grazing systems. Native herbivores can also have similar impacts on vegetation and wildfire [3,4], but specific behaviors can also increase wildfire risks [4]. An important distinction between grazing by wild and domestic herbivores on private and public lands is the ability to manage grazing in order to achieve specific vegetation management objectives. This review is focused on planned and managed herbivory, which is often not possible with wild herbivores and is therefore not discussed. Utilizing and manipulating livestock grazing for wildfire fuel management can be a sustainable alternative to other vegetation management methods when applied with an understanding of fire behavior, the forage environment and ecological objectives.

2. Concepts of fuel management

The intensity of wildfires is determined by thermal dynamics or the transfer of heat. Fuels must be preheated until absent of moisture and then it produces flammable gases that are easily ignited. The smaller the diameter of the material, the less heat input required for it to dry, produce gas and ignite. Larger diameter fuels, due to size or mass, require more heat before gas is produced for ignition. This is why the rate of spread of a grass fire is much faster than a brush fire. The horizontal density and or space between plants (fuel sources) will impact the transfer of heat that is required for pre heating across the landscape. The vertical space between plants will also impact the heat transfer. Continuous fuel in that plain is called ladder fuel. A continuum of fuel is one of the factors that controls flame height. Other factors that contribute to the fire behavior are the slope of the land surface and weather. A steeper slope will transfer heat between fuels more efficiently and create an explosive environment. In steep canyons, as the heat rises above to plants the angle combines horizontal and part of the vertical heat transfer. This is why most fuel reduction is conducted on flat topography areas like the tops of ridges.

Fuel treatments are generally arranged in two different approaches. Fuel breaks are linear fuel modifications often situated along a road or ridge. They can range in width from 10 to 120 meters and are designed as a tool for fire fighters to stop fires. Landscape area treatments are designed to reduce flame height and change fire behavior over a large area. Long term landscape treatment efforts are focused on changing the plant community to decrease the flame height when fire occurs. Both approaches require maintenance to remain valuable fire management tools. The objective for fuel reduction is to change fire behavior by impacting the following: fuel bed depth, fuel loading, percent cover, and ladder fuels that results in a fire flame of less than four feet. At that level all fire fighting management tools can be used while maintaining fire fighter safety.

3. Disturbance to reduce fuels

Interruption or the disturbance of the plant growth can be achieved through grazing, burning or other treatments. Mechanized disturbance treatments are used by land managers to alter or remove vegetation included mowing, mastication, and biomass harvesting. Mastication involves the use of a large mechanized device that chops shrubs and trees to break up the fuel pattern and decrease combustibility by placing fuels on the ground. It changes fire behavior by rearranging the fuel profile and by distributing some of the fuel on the ground. This action also causes a reduction of ladder fuels, which decreases potential for vertical extension of fire into tree canopies; crown fires are extremely difficult for fire fighters to control.

Mastication can be used as a pretreatment followed by prescribed fire or grazing treatments. Some of the disadvantages of mastication are the costs, ground disturbance, short life of the treatment in some areas, terrain and surface roughness limitations, and soil compaction. Mastication will result in death in some brush species, but many will re-sprout from the roots and require retreatment. Mechanized disturbance treatments also include the thinning of over-story vegetation through biomass harvesting. The harvested biomass is brought to a chipping unit and the resulting material is transported off the site for use in energy power plants. The sale of the biomass chips reduces the cost of this treatment. Thinning can provide desired conditions for both ladder fuels and crown spacing in one treatment. Soil moisture condition is the only limitation on the time of year that the treatment can be conducted. Disadvantages include transportation costs of hauling biomass and removal of nutrients from the ecosystem. In some cases, trees that are removed can be sold as commercial saw logs to offset fuel treatment costs.

Mowing is generally used in grass communities to drop the fuel on the ground, where it has less contact with air and decreases the combustibility. Mowing needs to be applied during end of the green season since it can cause fires from the blades striking rocks when dry grass is present.

Herbicides can be sprayed to kill specific plants, but this does not alter the fuel pattern immediately. Herbicide treatment of targeted species can be the cheapest methods. The disadvantages include concerns about its impact on the environment and short term increases in fuel flammability.

Prescribed fire can be used to change the fuel load and pattern. Air quality concerns and the need for the correct fire weather conditions (wind, air and plant humidity) may limit the use of prescribed fire to a narrow time period in the season that implementation can occur. A mechanical or hand removal treatment may also be required prior to the reintroduction of fire into the ecosystem to achieve desired fire behavior. The disadvantages of this treatment are reduced aesthetics, tree mortality, impaired air quality, liability concerns, pretreatment costs where applicable, required qualified people that understand prescribed fire, treatment variation (it may burn hotter or cooler than planned), and it may not be appropriate for some plant communities such as low-elevation sagebrush that can be converted to cheatgrass post fire.

Hand cutting and stacking of fuels for burning is very selective and is often the preferred method to treat larger diameter fuels on steep slopes where mechanized equipment cannot operate. The cost for this labor intensive method is comparatively high and depends on the type and amount of vegetation and terrain.

3.1. Grazing for fuel management

Grazing is best used when addressing the smaller diameter vegetation that make up the 1 and 10-hour fuels. One-hour fuels are those fuels with a moisture content that reaches equilibrium with the surrounding atmosphere within one hour and are less than 6 mm in diameter. Ten-hour fuels range from 6 to 25 mm in diameter. Grazing can impact the amount and arrangement of these fuels by ingestion or trampling as seen in Figure 1.

Figure 1. Goats altering the fine fuels.

Grazing is a complex dynamic tool with many plant and animal variables, which requires sufficient knowledge of the critical control points to reach treatment objectives. Those control points involve the species of livestock grazed (cattle, sheep, goats or a combination), the animals' previous grazing experience that will effect their preference for certain plants, time of year as it relates to plant physiology (as the animals consumption is directed by the seasonal nutrient content), the animal concentration or stocking density during grazing, grazing duration, plant secondary compounds, and animal physiological state. Grazing treatments can be a short term application to reduce flammable vegetation or a long term practice designed to change vegetation structure and composition through the depletion of root carbohydrates in perennials and the seed bank of annual plants. The fire prevention objectives are to change the fire behavior through modification of the fuel bed, fuel loading, percent cover, and ladder fuels.

The plant community and fire prevention objectives will determine the targeted vegetation of concern and the plants' life cycle (annual or perennial) will determine the type of grazing that will be applied for fuel management. Control of annual plants will require annual treatments that will remove plant material prior to the fire season. Grazing before seed set can change seed bank dynamics and long-term implementation of grazing can change plant species composition. Control of perennial plants will require repeated grazing treatments that deplete root carbohydrates and cause mortality of targeted species, which also changes plant species composition. Root carbohydrate reserves are at their lowest level just after the period when plants initiate active shoot elongation. If plants are severely grazed early in the growing season, carbohydrate reserves will be depleted and plant vigor reduced [5]. Removal of bark or repeated defoliation are two other ways to destroy perennial plants. In shrub species, the concept of changing the fuel profile the first year and managing it thereafter with grazing over large areas appears to be most sustainable.

Integration of different treatments could provide the best strategy. Livestock cannot effectively control mature shrubs that either grows higher than the animals can effectively graze or have large diameter limbs. Mastication, under burning, hand cutting can be used to manipulate the large diameter 100-hour shrub fuels and grazing can be used as a follow up treatment for controlling re-sprouting species or shifting the species composition to herbaceous plant fuel material. Tsiouvaras [6] suggests that grazing followed with prescribed fire can be used safely to kill the above ground part of shrubs and further open the stand. Magadlela [7] reported that cutting and herbicide increased sheep effectiveness by reducing the shrubs below 20% in one year, but increased the costs.

4. Grazing impacts on fuels

Prescribed grazing has the potential to be an ecologically and economically sustainable management tool for reduction of fuel loads. However, much of the information on grazing for fuel reduction is anecdotal and scientific research is limited. Existing data indicate there are two ways in which grazing impacts the fuel load, removal of vegetation and hoof incorporation of fine fuels. Smith *et al.* [8] found that in Nevada 350 ewes grazed intensively on *Artemisia tridentata* (sagebrush) and *Bromus tectorum* (cheatgrass) in a 2.5-mile fuel break divided into 20 pastures reduced fine fuels from 2,937 to 857 kg/hectare. Vegetative ground cover decreased 28 to 30 %, ground litter increased 20 to 23 % and bare ground increased 4%. Planned herbivory treatments in Idaho reduced cheatgrass biomass resulting in reductions in flame length and rate of spread. When the grazing treatments were repeated on the same plots in May 2006, cheatgrass biomass and cover were reduced to the point that fires did not carry in the grazed plots [9]. Tsiouvaras [6] studied grazing on a fuel break in a California *Pinus radiata* (Monterey pine) and eucalyptus forest in the fall at a stocking rate of 279 Spanish goats/hectare for three days and reduced the brush understory by 46% and 82% at a 58 centimeter and 150 centimeter height respectively. Forage biomass utilization by the goats in the brush understory was 84%. *Rubus ursinus* (California blackberry) showed the largest decrease in cover (73.5%) followed by *Heteromeles arbutifolia* (Toyon), *Baccharis pitularis* (coyote brush), *Lonicera spp.* (honeysuckle), herbaceous plants, and *Arbutus menziesii* (madrone). *Toxicodendron diversilobum* (poison oak and eucalyptus exhibited very little change. Grazing of goats not only broke up the sequence of live fuels, horizontally and vertically up to 150 centimeters, but also reduced the amount of 1 and 10-hour dead fuels 33.2% and 58.3% respectively, while the 100-hour fuels remained constant. The litter depth was also reduced as much as 27.4% (from 7.4 centimeters before to 5.1 centimeters after grazing). Animal trampling resulted in crushing of the fine fuels and mixing them into the mineral soil, thus reducing the chance of ignition. In Southern California Green *et al.* [10] grazed 400 goats to create fuel breaks through chaparral in July. The goats utilized 95% of the leaves and small twigs to 1.6 mm diameter from all the *Cercocarpus spp.* (mountain mahogany) plants. Use of *Quercus berberidifolia* (scrub oak) was 80%, while use of *Adenostoma fasciculatum* (chamise), *Arctostaphylos glandulosa campbelliae* (eastwood manzanita), and *Eriogonum fasciculatum foliolosum* (California buckwheat) was low and *Ceanothus spp.* was only taken under duress. Under "holding pen" conditions, use of less palatable species approached the use of palatable plants [10]. Lindler [11] reported that goats stocked at seven per acres for three weeks in the summer in a ponderosa pine forest had an estimated vegetation removal of 15 to 25% depending on the plant species present and the length of stay in the pasture. The cost of the grazing treatment was $148 to $173 per hectare. Herbicide comparison costs on adjacent sites were $148 to 309 per hectare and removed 75 to 90% of the vegetation understory in the pine forest. Intensive grazing of cattle to control shrub growth has been demonstrated as being useful that could be used for maintenance of fuel breaks [12-16].

Perevolotsky [17] found that mechanical shrub removal and cattle grazing at the peak of green season in Israel during four consecutive years proved the most effective firebreak treatment. Heavy grazing for a short duration removed more than 80% of the herbaceous biomass, but affected the regeneration rate of shrubs for only 2 years. They stated that using goats or other browsing animals may increase the amount of shrub material removed by direct grazing, but may decrease the physical damage to shrubs. Henkin [15] found that under heavy grazing (175–205 cow grazing days per hectare), the basal regrowth of the oaks was closely cropped and the vegetation was maintained as predominantly open woodland. In the paddock that was grazed more moderately (121–148 cow grazing days per hectare), the vegetation tended to return to dense thicket [15].

Each species of animal has a unique grazing utilization pattern that is a function of mouth size and design, past grazing experience, and optimization of nutritional needs [18]. The mouth size will control how closely animals are able to select and graze to a given surface. Animals also differ in their forage preferences and diet composition, thus when developing a fuel reduction grazing program it is important to select the type of livestock that will consume the desired species and alter the fire behavior. Provenza & Malechek [19] showed a 50% reduction of tannin in goat masticated samples compared to un-masticated samples. This illustrates the goats can tolerate one of the secondary compounds that are present in some shrub species allowing higher amounts intakes. When preferred forage is absent or unpalatable, grazing animals are capable of changing their food habitat.

Forage type	Animal species		
	Cattle	Sheep	Goats
Grass	78	53	50
Forbs	21	24	29
Browse	1	23	21

Table 1. Percent of time (%) spent by animals feeding on diverse plant types in Texas [20].

Magadlela [7] found that goats grazing in Appalachian shrubs defoliated shrubs early and then grazed herbaceous material later in the season. Sheep preferred to graze herbaceous material first, but increased grazing pressure forced sheep to defoliate shrubs earlier in the season. They found that goats reduced shrub cover from 45% to 15% in one year. Sheep took three years to create the same results. Goats had improved shrub clearing when they followed sheep, reducing total shrubs from 41 to 8% in one year. By the end of five years of goat grazing, the shrubs were reduced to 2% cover. Luginbuhl *et al.* [21] found that *Rosa multiflora* (multiflora rose) was practically eliminated from the Appalachian Mountains after four years of grazing by goats alone (100%) or goat + cattle (92%). Simultaneously, vegetative cover was increased with only goats (65 to 86%) and with goats + cattle (65 to 80%), compared with the control plot where vegetation cover decreased from 70 to 22%. Lombardi *et al.* [22] studied the use of horses, cattle and sheep in Northwest Italy for five years and found that grazing reduced woody species cover and stopped the expansion of shrub population. The impact varied with animal. Cattle and horses had a higher impact on the plants caused by trampling. They found that the effectiveness of control depended on palatability and tolerance of woody species to repeated disturbance. Juniper and Rhododendron species were reported not to have been grazed. Hadar *et al.* [16] reported that the inconsistent response of some plants to grazing could be the interaction between grazing pressure and moisture conditions. They found that heavy cattle grazing (840 - 973 cow grazing days per hectare) during 7 to 14 days at the end of the growing season decreased species richness by consuming the seeds of herbaceous plants.

Sheep and goats grazing California chaparral presented dissimilar foraging strategies over the three grazing seasons [23]. They selected fairly similar species, but in different proportions at different seasons. Narvaez [23] found the proportion of browse in sheep and goat diets was greater when shrubs in chaparral areas were more abundant than herbaceous species. Browse accounted for 86.7% of the total forage ingested by goats and 71% by sheep. Seasonal grazing differences were also observed with sheep shifting from a browse dominated diet in fall and winter months to an herbaceous dominated diet in the spring when grasses were abundant and at their most nutritious state for the year. Goats maintained a browsing preference across all seasons and had a higher dry matter and nutrient intake than sheep over the three grazing seasons. Dry matter intake for goats was sufficient to meet maintenance requirements as was not the case with sheep. Goats were more effective than sheep in reducing fuel load in California chaparral as they consumed more vegetation and did not appear to be nutritionally limited by the low quality of the landscape. Sheep may be more effective in an herbaceous dominated landscape for fuel load reduction.

The impact of grazing on specific plant species will depend on the time of year grazing is applied. Herbivores will respond to the nutritional status of plants and their parts by selecting and concentrating their consumption on the most palatable and nutritious parts. As the physiological status of a plant changes throughout the year, the nutritional value of its parts change which can increase or decrease the desirability of those parts to herbivores. Taylor [20] reported studies in Idaho using heavy grazing by sheep showed that season of use impacted the utilization. Late-fall grazing reduced *Artemisia tripartita* (three tip sagebrush), while grazing during the spring increased sagebrush and decreased grasses.

Grazing impact can change with the density of animals and duration of grazing. The shorter the duration the more even the plain of nutrition will be. Over long periods of time in a pasture animals will first select the most nutritious forage and then move down in their preference of plants consumed. Stock density will have a great impact on the consumption and trampling of fuels. Fences, herding, topography, slope, aspect, distance from water, placement of salt, and forage density will all impact the distribution of animals and their

Figure 2. Electric fence netting for targeted goat grazing.

utilization of the forage. By concentrating animals into a smaller area for short periods of time, plant preference and selectivity will decrease as animals compete for the available forage. Increasing stock density will also increase hoof action and incorporation of the fine fuels into the ground. Spurlock *et al.* [24] stated that high stocking rates with little supplementation forces goats to graze even less palatable species and plant parts and resulting in the eradication of many shrubs in 2-3 years. Lindler [11] suggests that a stocking rate of 37 goats per hectare in a California pine forest is required to effectively treat understory brush.

Stoking rate	Forage type		
	Browse	Grass	Forbs
Light	16	55	28
Heavy	55	39	5

Table 2. Sheep diet consumption in Texas varied with stocking rate [25].

Grazing intensity	Bare soil	Vegetation cover (%)	Litter
Light	+6	-22	+25
Moderate	+4	-28	+20
Heavy	+4	-30	+23

Table 3. Results with sagebrush/grass pastures grazed at different intensities by sheep in northern Nevada [8].

Hadar [16] reported that light grazing provided greater plant diversity on treated sites. Thus, when proposing a stocking rate for treatment consumption, the environmental impact needs to be considered.

5. Nutritional and anti-nutritional factors

Low nutritional value and the presence of secondary compounds, such as tannins, in many California chaparral species are limiting factors for their use as forage by animals grazing this type of vegetation, especially during summer and fall [23]. The most abundant California chaparral species had low crude protein content (< 8%) and low digestibility especially in the summer and fall. This would include: *Adenostoma fasciculatum* (chamise), *Arctostaphylos canescens* (hoary manzanita), *Arctostaphylos glandulosa* (Eastwood manzanita), *Arctostaphylos stanfordiana* (Stanford manzanita), *Baccharis pitularis* (coyote brush), *Ceanothus cuneatus* (buck brush), *Eriodictyon californicum* (yerba santa), *Quercus durata* (leather oak), *Heteromeles arbutifola* (toyon), *Quercus douglasii* (blue oak), and *Quercus wislizenii* (interior live oak). Chaparral plants with the highest crude protein from leaf and stem samples included: *Baccharis pitularis* (coyote brush), *Ceanothus cuneatus* (buck brush), and *Eriodictyon californicum* (yerba santa) [23].

Ruminant diets with crude protein below 7-8% reduce feed intake because it does not provide the minimum rumen ammonia concentration for microbial growth. Nutritional supplementation would be needed for optimum performance in small ruminants used to reduce fuel loads in California chaparral. California chaparral had high fiber (neutral detergent fiber, NDF and acid detergent fiber, ADF) in most shrubs. *Baccharis pitularis* (coyote brush) and *Eriodictyon californicum* (yerba santa) had the lowest fiber concentrations. Organic matter digestibility and metabolizable energy were higher during spring plant growth for all species tested [23]. Taylor found that cottonseed meal and alfalfa supplements increased redberry juniper consumption by 40% [26].

Over time plants have developed mechanisms to limit or prohibit herbivory. Launchbaugh *et al.* [27] summarized this plant-animal interaction as follows: plants possess a wide variety of compounds and growth forms that are termed "anti-quality" factors because they reduce forage's digestible nutrients and energy or yield a toxic effect that deter grazing. Secondary compounds (e.g. tannins, alkaloids, oxalates, terpenes) can control the plant-animal interactions that drive intake and selection.

California chaparral plants with the highest total condensed tannins include: *Arctostaphylos canescens* (hoary manzanita), *Arctostaphylos glandulosa* (Eastwood manzanita), *Arctostaphylos stanfordiana* (Stanford manzanita), *Ceanothus cuneatus* (buck brush), and *Quercus douglasii* (blue oak). Narvaez [23] showed that condensed tannins concentrations in California chaparral shrubs might negatively impact ruminant feed utilization in addition to the impact of protein binding.

Forage intake and digestibility of two common chaparral shrubs, *Adenostoma fasciculatum* (chamise) and *Quercus douglasii* (blue oak), as a sole diet were low and did not meet the nutritional requirements for sheep and goats grazing in this type of vegetation [23]. Greater understanding of nutrition of chaparral shrubs being grazed in prescribed herbivory and monitoring of animal condition are needed to know when and what to use for strategic supplementation or replace thin animals with those in better condition.

Animals may expel toxic plant material quickly after ingestion, secrete substances in the mouth or gut to render the compounds inert, or rely on the rumen microbes or the body to detoxify them. The grazing practitioner can address plant toxins in different ways. A species of livestock can be selected that can detoxify compounds or have a smaller mouth that allows them to eat around thorns. Nutritional or pharmaceutical products can be offered to aid in digestion and detoxification. Breeding stock can be selected based on an individual animal's tolerance to toxic compounds. Tannins are the most important defense compounds present in browse, shrubs, and legumes forages. Concentrations in woody species vary with environment, season, plant developmental phase, plant physiological age, and plant part. Levels in excess of 50 g/kg DM can reduce palatability, digestibility, voluntary feed intake and digestive enzymatic activity and can be toxic to rumen micro-organisms [28-32]. In some cases, when the plant compound is known, methods of interceding can be used. Polyethylene glycol (PEG) is a polymer that binds tannins irreversibly, reducing the negative effects of tannins on food intake, digestibility, and preferences [33]. Polyethelyne glycol was used in California to overcome the protein binding of tannins and make protein and energy more available to sheep and goats. Supplementation with PEG significantly increased consumption of *Arctostaphylos. canescens* (hoary manzanita) by small ruminants [23]. Appropriate nutritional and non-nutritional supplementation may help develop prescribed herbivory into a viable fire fuel management strategy for California and other areas with chaparral plant communities. More nutritional analysis of shrubs and increased understanding of the impact of associated plant secondary compounds on consumption and utilization by ruminants are needed.

For oxalates, calcium supplementation has shown to ameliorate the diet suppression. Launchbaugh [27] suggested that supplementation of protein, phosphorous, sulfur, and energy can also make a difference in intake of plant material containing secondary compounds. They even postulate that clay could be used to detoxify compounds.

6. Integrating grazing into the ecosystem

It is important to recognize the different viewpoints people will have on using grazing for vegetation management purposes. These viewpoints can affect the way grazing is applied, the long-term success of grazing for controlling wildfire fuels and the cost of using grazing. If grazing is viewed and used as another tool or method to be applied as other vegetation control methods (i.e. mechanical and chemical methods), the success may be limited and the cost of grazing may be greater than necessary. An alternative is a systems approach in which grazing is integrated as part of the ecosystem so that the system is both benefited by and benefits grazing.

Under a systems approach grazing becomes a more regular disturbance pattern that encourages growth of herbaceous vegetation and the smaller diameter fuels that are more nutritious and readily consumed by herbivores. These fuel classes are important as they can greatly impact the rate of spread of a fire along with the flame height. When grazing is used infrequently, as it often is when viewed in the same context as other single event fuel treatments, the vegetation will likely consist of older vegetation of poor nutrition that is more costly to graze due to the higher physiological cost to the animal and higher labor inputs for managing portable fencing. A regular grazing regime will create improved nutrition by providing smaller re-growth of higher nutrition vegetation allowing animal performance to improve while maintaining a desirable fuel profile. Weber *et al.* [34] found compelling evidence that regular livestock grazing on public land grazing allotments between the years 1993 and 2000 effectively maintained a lower fuel profile and reduced the risk of wildfires.

Figure 3. Goats grazing a treated ridge following other land treatments.

Another aspect of a systems approach to managing wildfire fuels with grazing is to strategically use grazing in combination with other methods of vegetation management. Weber *et al.* [34] found that wildfire and grazing alone reduced mean fuel loads 38% and 47% respectively compared to control treatments. When the effects of wildfire and grazing were combined fuel loads were reduced 53%. Integrating fire and grazing in a strategic manner can provide conservation benefits and increase livestock performance. In an 11-year study pyric-herbivory, or patch burning, was applied to tallgrass and mixed-grass prairie in the United States to re-introduce more natural fire regimes and improve wildlife habitat [35]. Livestock performance was not affected by the use of pyric-herbivory on the tallgrass prairie (8 years) while on the mixed-grass prairie stocker cattle had greater weight gains and more consistent performance over the 11-year period [35]. Another successful combination of vegetation management methods that is often employed in areas with larger diameter woody fuels is to initially use mechanical treatments to reduce the woody biomass and then apply grazing to maintain a shorter and more herbaceous vegetation structure. The combination of vegetation control methods in managing wildfire fuels is consistent with the Integrated Pest Management (IPM) strategies commonly and successfully used in agricultural pest management systems.

Grazing is best used when addressing the smaller diameter vegetation that make up the 1 and 10-hour fuels. These two fuel classes are important as they can greatly impact the rate of spread of a fire along with the flame height. Many fire managers have looked at grazing in the same context as other single event mechanical fuel treatments. These grazing treatments have been expensive to implement as they have a physiological cost to the animal and higher costs of portable fencing to reach fuel objectives in one year. Perhaps a sustainable use of grazing would be annual grazing of large areas following mechanical treatment. This will provide improved nutrition by providing smaller regrowth that is higher in nutrition allowing animal performance to improve while maintaining a specific fuel profile.

7. Practical considerations

Grazing animals can effectively distinguish between plants that differ in digestible energy or nutrients. The animal's consumption is driven by their physiological state. Non-lactating animals have much lower nutrient requirements than lactating females or growing weaned animals and can consume a wider array of plants to meet their nutritional needs. Animals can be forced to eat below their nutritional needs and they will balance their needs by catabolizing body fat and protein. The animal can tolerate short-term energy or protein deficits, but sustained periods at this status can be reason for concern. For this reason lactating and young growing animals may not be recommended for fuel control. Growing animals can be used to consume new shrub growth in a shrub grazing system designed to maintain the fuel profile.

Because of the complexity of plant and animal interactions, a project evaluation should be developed considering measurable and attainable objectives before grazing is used. It should include a review of treatment objectives, desired outcomes, and environmental impacts. This will dictate the kind of animal needed, grazing intensity, timing of the grazing event, and duration of the grazing period. Variation in animal-plant interaction is driven by forage type, grazing season, yearly season variation, animal interaction with the grazing system (animal density and competition), previous grazing experience, mixture of grazing animals, and pre-grazing treatment (integrated approach). The treatment and resulting outcomes are not conveniently predicted and may require adaptive onsite management. Treatment standards include stubble height for grass, percent vegetation cover by shrubs, plant mortality, or removal of 1 and 10-hour fuel and fuel bed depth.

Any grazing plan designed for fuel reduction will need to review the grazing impacts on parameters other than just fuel reduction. The effects of the grazing management should be studied for its impact on water quality, soil compaction, riparian vegetation, disease exposure from wildlife (bluetongue, pasturella) and weed transmission. The positive aspects of grazing over other treatments should also be weighed, including the recycling of nutrients into the products of food and fiber.

The grazing contractor will use, in most cases, portable electric polywire or netting to contain small ruminants in an area. A low-impedance solar-powered energizer with adequate grounding will power the electric fence material. Predators will be a concern for small ruminant safety and will require use of a guardian animal for protection. Guardian dogs are the preferred choice in most remote areas. Herders may be needed on large contracts. Mineral supplementation will be necessary to keep animals productive and healthy. Protein supplements may be needed in fall and summer. Lack of available stock water will require a way to haul water to meet daily requirements. In hot weather, water intake of small ruminants can approach two gallons per head per day. A truck and trailer will be needed to haul animals and a herding dog will most likely be needed for moving stock. Adequate general liability and automobile insurance will be required in bids and must be maintained by the contractor. Livestock and full mortality insurance should be considered. Third party firefighting and fire suppression expense liability should be considered if doing many fuel load reduction or firebreak contract.

The social aspect is often an important and overlooked part of prescribed herbivory in contract grazing situations. Grazing contractors will benefit by taking the time to engage the general public in explaining and answering questions regarding grazing and animal husbandry. Suburban and urban residents commonly question concerns about perceived loss of wildlife habitat or landscape view, guardian animals and animal welfare when new grazing projects are implemented adjacent to populated areas. These topics need to be addressed in a calm, rational manner. Timely corrective response to any issues such as livestock escaping fences will be important.

Current and historical perceptions by the public of grazing will influence acceptance and understanding of grazing treatments for fuel control. It is important for grazing contractors to have well defined contracts and consider public education as one of their roles, especially with contracts near residential areas. Consumptive use, such as grazing, may not be compatible with recreation land use in some areas. A survey by Smith *et al.* [8] indicated that 90% of residents near a fuel break stated that sheep were a preferred alternative for fuel reduction. Only 10% felt that they were inconvenienced by the treatment. Some responses indicated the ignorance of many residents to grazing and grazing management, such as concerns of electrocution of animals and humans by the electric fence. This condition will need to be addressed when making grazing proposals with an understood that public education will be a necessary part of the process.

8. Conclusion

The modification of wildfire fuels is an important issue in many regions of the world. The use of grazing animals for fuel management has a limited research knowledge base to direct the timing and intensity to reach the fuel management objectives in comparison to other methods. Also seasonal variation of nutrition content and secondary compounds of shrubs need to be further defined. Most of the grazing fuel modification study work has been conducted with goats, due to their preference for targeted plant species. Grazing animals can modify wildfire fuels through consumption and trampling. Animals are most effective at treating smaller diameter live fuels and 1and 10-hour dead fuels. These fuels are important components of fire behavior by providing the flammable material that creates a ladder of fuel for a fire to extend up from the ground into the shrub and tree canopy. Science-based research on the use of grazing to achieve fuel management objectives exists, but is very limited and many studies only had a single-year grazing treatment. In a grass ecosystem this may be effective if timed correctly, but shrub vegetation often require grazing treatments over multiple years to create and maintain a fuel profile that is more desirable.

There are many issues that need to be considered as part of grazing for fuel reduction. Grazing has a more varied outcome than the mechanical fuel reduction treatments. Until grazing is viewed in a systems approach in which the numerous factors that affect grazing effectiveness are considered, the dominant management will be to force utilization by limiting nutrition and or preference. The understanding of animal preference and the proper timing and livestock management required to meet the objective are all critical elements in implementing an effective and sustainable grazing program for wildfire fuel management.

Author details

Roger S. Ingram*
University of California Cooperative Extension, Auburn, California, United States

Morgan P. Doran
University of California Cooperative Extension, Fairfield, California, United States

Glenn Nader
University of California Cooperative Extension, Yuba City, California, United States

Acknowledgement

The authors thank Ed Smith of the University of Nevada Cooperative Extension, Zalmen Henkin of the Agricultural Research Organization at Bet Dagan, Israel, Nelmy Narvaez, past graduate student at University of California, Davis and Wolfgang Pittroff for their contributions. Bill Burrows and the Sunflower CRMP provided the pictures.

9. References

[1] National Interagency Fire Center. Wildland Fire Statistics 2006. http://www.nifc.gov/fireInfo/fireInfo_stats_YTD2006.html (accessed 24 April 2012)

[2] Davidson J. Livestock grazing in wildland fuel management programs. Rangelands 1996; 18(6) 242-245.

[3] Risenhoover KL. and S.A. Maass. The influence of moose on the composition and structure of Isle Royale forests. Canadian Journal of Forest Research 1987;17 357-364.

[4] Hierro JL, Clark KL, Branch LC, Villarreal D. Native herbivore exerts contrasting effects on fire regime and vegetation structure. Oecologia 2011;166 1121-1129.

[5] Doescher PS, Tesch SD, Alejandro-Castro M. Livestock grazing: A silvicultural tool for plantation establishment. Journal of Forestry 1987;85 29-37.

* Corresponding Author

[6] Tsiouvaras CN, Havlik NA, Bartolome JW. Effects of goats on understory vegetation and fire hazard reduction in a coastal forest in California. Forest Science 1989;35 1125-1131.

[7] Magadlea AM, Dabaan ME, Bryan WB, Prigge EC. Brush clearing on hill land pasture with sheep and goats. Journal of Agronomy and Crop Science 1995;174 1-8.

[8] Smith E, Davidson J, Glimp H. Controlled Sheep grazing to create fuelbreaks along the urban-wildland interface. In: proceedings of the Society of Range Management 53rd Annual meeting,13-18 February 2000, Boise, Idaho 2000.

[9] Diamond JM, Call CA, Devoe N. Effects of targeted cattle grazing on fire behavior of cheatgrass-dominated rangeland in the northern Great Basin, USA. International Journal of Wildland Fire 2009;18 944–950. http://www.publish.csiro.au/?paper=WF08075 (accessed 24 April 2012)

[10] Green LR, Hughes CL, Graves WL. Goats control of brush regrowth on Southern California fuel-breaks. In: 1st International Rangeland Congress, 14-18 August 1987, Denver, Colorado; 1987.

[11] Lindler D, Warshawer J, Campos D. Using goats to control understory vegetation. In: proceedings of the 20th Forest Vegetation Management Conference, 19-21 January 1999, Redding, California. 1999.

[12] Allen BH, Bartolome JW. Cattle grazing effects on understory cover and tree growth in mixed conifer clearcuts. Northwest Science 1989;63 (5) 214-220.

[13] Gutman M, Henkin Z, Holzer Z, Noy-Meir I, Seligman NG. Beef cattle grazing to create firebreaks in a Mediterranean oak maquis in Israel. In: Proceedings of the IV International Rangeland Congress, 22-26 April 1991, Montpellier, France. 1991.

[14] Masson P, Guisset C. Herbaceous and shrubby vegetation evolution in grazed cork oak forest firebreaks sown with subterranean clover and perennial grasses. In: Proceedings of the 7th Meeting FAO European Sub-network on Mediterranean Pastures and Fodder Crops, 21-23 April 1993, Crete, Greece. 1993

[15] Henkin Z, Gutman M, Aharon H, Perevolotsky A, Ungar ED, Seligman NG. Suitability of Mediterranean oak woodland for beef herd husbandry. Agriculture, Ecosystems and Environment 2005;109 255–261.

[16] Hadar L, Noy-Meir I, Perevolostsky A. The effect of shrub clearing and intensive grazing on the composition of a mediterranean plant community at the functional group and species level. Journal of Vegetation Science. 1999;10 673-682.

[17] Perevolotsky A, Schwartz-Tzachor R, Yonatan R. Management of fuel breaks in the Israeli mediterranean ecosystem. Journal of Mediterranean Ecology 2002;3(2-3) 13-22.

[18] Beasom SL. Dietary overlap between cattle, domestic sheep, and pronghorns. In Soesbee, Ronald E; Guthery, Fred S. eds. Noxious Brush and Weed Control Research highlights-. Lubbock, Tx: Texas Tech. University; 1980;11 40-41.

[19] Provenza FD, Malechek JC. Diet selection by domestic goats in relation to blackbush twig chemistry. Journal of Applied Ecology 1984; 21:831–84

[20] Taylor CA. Sheep grazing as a brush and fine fuel management tool. Sheep Research Journal 1994;10(1) 92:96

[21] Luginbuhl JM, Harvey TE, Green JT, Poore MH, Mueller JP. Use of goats as a biological agents for the renovation of pastures in the Appalachian region of the United States. Agroforestry Systems 1999:44 241-252.

[22] Lombardi G, Reyneri A, Cavallero A. Grazing animals controlling woody-species encroachment in subalpine grasslands. In: Proceedings of the International Occasional Symposium of the European Grassland Federation, 27-29 May 1999, Thessaloniki, Greece. 1999.

[23] Narvaez N. Prescribed herbivory to reduce fuel load in California chaparral. PhD thesis, University of California, Davis; 2007.

[24] Spurlock GM, Plaister R, Graves WL, Adams TE, Bushnell R. Goats for California brushland. Cooperative Agriculture Extension, University of California. 1980; Leaflet 21044.

[25] Kothmann MM. The botanical composition and nutrient content of the diet of sheep grazing on poor condition pasture compared to good condition pasture. Ph.D dissertation. Texas A&M University, College Station; 1968.

[26] Taylor CA, Launchbaugh KL, Huston JE, Straka EJ. Improving the efficacy of goating for biological juniper management. In: 2001 Juniper Symposium. Texas A&M University Research and Extension Center, San Angelo, Texas. 2001.

[27] Launchbaugh KL, Provenza FD, Pfister JA. Herbivore response to anti-quality factors in forages. Journal Range Management 2001;54 431-440.

[28] Robbins H, Hagerman AE, Hajeljord O, Baker DL, Schwartz CC, Moutz WW. Role of tannins in defending plants against ruminants: reduction in protein availability. Ecology 1987;68 98-107.

[29] Happe PJ, Jenkins KJ, Starkey EE, Sharrow SH. Nutritional quality and tannin astringency of browse in clear-cuts and old-growth forest. Journal Wildlife Management 1990;54 557-556.

[30] Kumar R, Vaithyanathan S. Occurrence nutritional significance and effect on animal productivity of tannins in tree leaves. Animal Feed Science Technology 1990;30 21-38.

[31] Lowry JB, McSweeney CS, Palmer B. Changing perceptions of the effect of plant phenolics on nutrient supply in the ruminant. Australian. Journal of Agriculture Research 1996;47 829-842.

[32] Bryant JP, Provenza FD, Pastor J, Reichardt PB, Clausen TP, du Toit JT. Interactions between woody plants and browsing mammals mediated by secondary metabolites. *Annual Review of Ecology and Systematics* 1991;22 431-446.

[33] Villalba JJ, Provenza FD, Banner RE. Influence of macronutrients and polyethylene glycol on intake of a quebracho tannin diet by sheep and goats. Journal Animal Science 2002;80 3154-3164.

[34] Weber KT, McMahan B, Russell G. Effect of Livestock Grazing and Fire History on Fuel Load in Sagebrush-Steppe Rangelands - In: Wildfire Effects on Rangeland Ecosystems and Livestock Grazing in Idaho. 2011. http://giscenter.isu.edu/research/techpg/nasa_wildfire/Final_Report/Documents/Chapter9.pdf (accessed 24 April 2012)

[35] Limb RF, Fuhlendorf SD, Engle DM, Weir JR, Elmore RD, Bidwell TG. Pyric-herbivory and cattle performance in grassland ecosystems. Rangeland Ecology and Management 2011;64 659-663